Sébastien Perseguers

Entanglement distribution in quantum networks

Sébastien Perseguers

Entanglement distribution in quantum networks

An investigation of the properties of communication networks based on the laws of quantum physics

Südwestdeutscher Verlag für Hochschulschriften

Impressum/Imprint (nur für Deutschland/ only for Germany)
Bibliografische Information der Deutschen Nationalbibliothek: Die Deutsche Nationalbibliothek verzeichnet diese Publikation in der Deutschen Nationalbibliografie; detaillierte bibliografische Daten sind im Internet über http://dnb.d-nb.de abrufbar.

Alle in diesem Buch genannten Marken und Produktnamen unterliegen warenzeichen-, marken- oder patentrechtlichem Schutz bzw. sind Warenzeichen oder eingetragene Warenzeichen der jeweiligen Inhaber. Die Wiedergabe von Marken, Produktnamen, Gebrauchsnamen, Handelsnamen, Warenbezeichnungen u.s.w. in diesem Werk berechtigt auch ohne besondere Kennzeichnung nicht zu der Annahme, dass solche Namen im Sinne der Warenzeichen- und Markenschutzgesetzgebung als frei zu betrachten wären und daher von jedermann benutzt werden dürften.

Verlag: Südwestdeutscher Verlag für Hochschulschriften Aktiengesellschaft & Co. KG
Dudweiler Landstr. 99, 66123 Saarbrücken, Deutschland
Telefon +49 681 37 20 271-1, Telefax +49 681 37 20 271-0
Email: info@svh-verlag.de
Zugl.: München, TU, Diss., 2010

Herstellung in Deutschland:
Schaltungsdienst Lange o.H.G., Berlin
Books on Demand GmbH, Norderstedt
Reha GmbH, Saarbrücken
Amazon Distribution GmbH, Leipzig
ISBN: 978-3-8381-1805-5

Imprint (only for USA, GB)
Bibliographic information published by the Deutsche Nationalbibliothek: The Deutsche Nationalbibliothek lists this publication in the Deutsche Nationalbibliografie; detailed bibliographic data are available in the Internet at http://dnb.d-nb.de.

Any brand names and product names mentioned in this book are subject to trademark, brand or patent protection and are trademarks or registered trademarks of their respective holders. The use of brand names, product names, common names, trade names, product descriptions etc. even without a particular marking in this works is in no way to be construed to mean that such names may be regarded as unrestricted in respect of trademark and brand protection legislation and could thus be used by anyone.

Publisher: Südwestdeutscher Verlag für Hochschulschriften Aktiengesellschaft & Co. KG
Dudweiler Landstr. 99, 66123 Saarbrücken, Germany
Phone +49 681 37 20 271-1, Fax +49 681 37 20 271-0
Email: info@svh-verlag.de

Printed in the U.S.A.
Printed in the U.K. by (see last page)
ISBN: 978-3-8381-1805-5

Copyright © 2010 by the author and Südwestdeutscher Verlag für Hochschulschriften Aktiengesellschaft & Co. KG and licensors
All rights reserved. Saarbrücken 2010

> *Every great and deep difficulty bears in itself its own solution. It forces us to change our thinking in order to find it.*
>
> — NIELS BOHR

Contents

Introduction   1

I   Pure states   7

1   Entanglement manipulation in basic networks   9
- 1.1   Entanglement swapping 10
  - 1.1.1   Joint measurement at the middle station 10
  - 1.1.2   Figures of merit 12
  - 1.1.3   Optimal measurements 13
- 1.2   Maximally-entangled states are not always optimum 16
  - 1.2.1   Two consecutive entanglement swappings 17
  - 1.2.2   A single square network 20
- 1.3   An infinite chain of quantum relays 22
  - 1.3.1   Exponential decay of the entanglement 23
  - 1.3.2   SCP under ZZ measurements 23

2   Long-distance entanglement in planar graphs   27
- 2.1   Deterministic methods 28
  - 2.1.1   Hierarchical graphs 28
  - 2.1.2   Regular lattices 31
- 2.2   Strategies based on bond percolation 32
  - 2.2.1   Classical entanglement percolation 34
  - 2.2.2   Quantum entanglement percolation 35
- 2.3   Multipartite entanglement percolation 39
  - 2.3.1   Generalized entanglement swapping 40
  - 2.3.2   An illustrative example 41
  - 2.3.3   The superiority of multipartite strategies 44
- 2.4   On optimal protocols 47

3   Quantum complex networks   51
- 3.1   The model 52
  - 3.1.1   Random graphs 52
  - 3.1.2   Erdős-Rényi networks in the quantum world 53
- 3.2   Joint measurements help 55
  - 3.2.1   Creation of W states 55

		3.2.2	Creation of GHZ states	56
	3.3	A complete collapse of the critical exponents		59
		3.3.1	The Λ subgraph	59
		3.3.2	General subgraphs	60

II Mixed states
63

4 Towards noisy quantum networks
65

	4.1	Rank-two mixed states		66
		4.1.1	From pure to mixed states and vice versa	66
		4.1.2	Quantum complex networks	68
	4.2	Full-rank mixed states		69
		4.2.1	Elementary operations on Werner states	69
		4.2.2	Quantum repeaters	71
		4.2.3	Lower bound for long-range entanglement	72

5 One-shot protocol in square lattices
73

	5.1	Network with bit-flip errors only		73
		5.1.1	Propagating a large GHZ state	74
		5.1.2	Network-based bit-flip error correction	75
	5.2	A fault-tolerant protocol via encoding		79
		5.2.1	Required physical and temporal resources	80
		5.2.2	Towards a realistic scenario	83

6 Fidelity threshold in cubic networks
87

	6.1	Quantum networks and cluster states		87
		6.1.1	A mapping to noisy cluster states	88
	6.2	Long-distance entanglement generation		90
		6.2.1	Measurement pattern and quantum correlations	90
		6.2.2	Error correction and fidelity of the final state	92
		6.2.3	Numerical estimation of the fidelity threshold	96

Bibliography
99

Acknowledgment
109

Introduction

> *Any sufficiently advanced technology is indistinguishable from magic.*
>
> — ARTHUR C. CLARKE

Nature is a perpetual source of wonder for mankind. It has been dazzling human beings since their very origin, and there is no doubt that this astonishment will never cease. However, before serving man, new phenomena have always been accompanied by incredulity and fear. Thousands of years of civilization taught us how to apprehend such discoveries, and mystical interpretations of Nature have been replaced by rational explanations. It is nevertheless a fact that man feels very uncomfortable when doubt is cast on his perception of the world he lives in. History has been shaken many times by scientific revolutions, and each improved the well-being, and hopefully the wisdom, of humanity. It is generally agreed that modern science finds its roots back in 1543 with the work of Nicolaus Copernicus, and since then many illustrious physicists, as Galileo Galilei, Isaac Newton or James C. Maxwell to name just a few, radically modified our conception of the universe. Acceptance of new theories is nonetheless a long-term process, not only for the general public but also among the experts in the field. Experimental observations of their predictions play a crucial role in that respect, and usually engineers close the discussion by making new and "magic" technologies out of them.

Quantum mechanics has revolutionized our daily lives, and it is no surprise that its confrontation with the general relativity of Albert Einstein and with the information theory of Claude E. Shannon, the two other major scientific achievements of the 20th century, has raised deep and fundamental questions about our world. The most famous example of the tension existing between these theories is certainly the Einstein-Podolsky-Rosen paradox, a *Gedankexperiment* arguing that quantum mechanics cannot be a complete and realistic physical theory [EPR35]. Some thirty years later, John S. Bell proved that there exist certain experimental settings that would contradict such a classical picture of reality [Bel64]. However, the debate among physicists remained passionate until the first experiments revealed the true nature of our world [FC72, FT76, AGR81, ADR82]: the statistical predictions of quantum mechanics, as originally formulated in the late 1920's, were confirmed. (Actually, there were some loopholes in the experiments, but a detailed discussion would bring us much too far from the scope of this introduction.)

Besides the fundamental and, to some extent, philosophical questions raised by quantum mechanics, some more pragmatic researchers saw in it a source of formidable possibilities, which initiated a "second quantum revolution" [Bel04, p. xix]. For instance, in contrast to classical encryption protocols, the possibility of using genuine quantum characteristics of light to achieve unconditionally secure communication between two distant parties was pointed out by Stephen Wiesner in the 1970's [Wie83]. This was rediscovered and popularized as *quantum cryptography* by Charles H. Bennett and Gilles Brassard [BB84]. In their work, they showed that a perfectly secure key distribution is possible by using quantum particles, since no one can eavesdrop without leaving a trace. This comes from one basic property of quantum physics which has been, and still is, a source of antagonistic interpretations: no measurement can be performed on a system without perturbing it. Some years later, Artur K. Ekert designed another scheme for quantum cryptography [Eke91]. In his Letter, he proposed to utilize the *quantum entanglement* ("spooky action at a distance" [Ein71]) that lies at the very heart of the EPR paradox and a generalized Bell theorem, known as the Clauser-Horner-Shimony-Holt inequalities [CHSH69], to test for eavesdropping. Quoting the introduction to [GRTZ02], these developments really show that the old and weird viewpoint considering quantum physics, due to its contrast with classical physics, as a set of negative rules stating things that cannot be done[1] was finally turned positive.

Entanglement turns out to be a wonderful resource for many communication protocols, such as superdense coding [BW92], quantum teleportation [BBC+93], or distributed quantum computation[2] [CEHM99], just to name a few. The implementation of these concepts requires an extraordinarily careful preparation of non-classical states, which are extremely fragile against any perturbation, but the control of physical systems at the quantum scale has been impressively increased over the last few years. Advanced technologies relying on entanglement are indeed being actively developed, and quantum key distribution systems in particular have matured to real-world application. For example, quantum cryptography was used to protect voting ballots casts in Geneva (Switzerland) during parliamentary elections on October 21, 2007. Another example is given by the network consisting of five stations and seven quantum connections that has been built recently in Vienna [PPM08], see Fig. 1a. This setup operates in a point-to-point modus, which means that the stations behave classically: quantum correlations are created between neighboring stations but cannot be transmitted farther. It is nonetheless very reasonable to predict that genuine *quantum networks* (Fig. 1b), that is, sets of stations sharing entangled pairs of particles that can be manipulated *ad libitum*, will appear in the near future.

Currently, remote entanglement is best created between atoms by sending single polarized photons, or beams of squeezed light, through optical fibers[3] [CZKM97, Kim08]. While local operations on the quantum systems are performed more and more reliably [MHPC06, HSP10],

---

[1]For example, one cannot determine both the position and the momentum of a particle with arbitrarily high accuracy, or duplicate an unknown quantum state.

[2]Quantum computation is another revolutionary field of quantum physics [Fey82, Deu85, Sho94, DiV95, Gro96]. However, the role played by entanglement in the quantum speed-up over classical computers is not clear at the moment [LBAW08, GFE09].

[3]A direct transmission of entangled photons over free-space links is also possible [FUH+09], but the relevance of this method highly depends on atmospheric conditions and thus may not be appropriate in hostile environments such as cities.

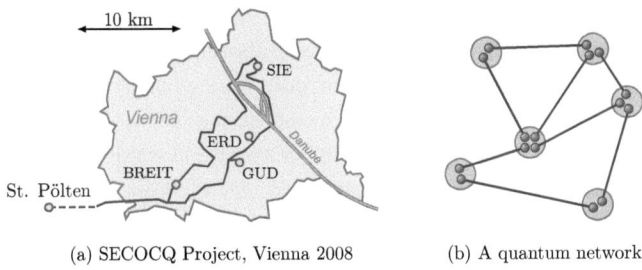

(a) SECOCQ Project, Vienna 2008   (b) A quantum network

Figure 1: (a) The fiber ring connecting four buildings in Vienna and one in St. Pölten, a city about 85 km distant from the station BREIT. (b) In general quantum networks, links represent entangled pairs of particles and any local quantum operations can be performed at the stations, which are located at the vertices of the graph.

one major challenge in quantum communication remains the generation of entanglement over a large distance. In fact, the imperfections of the quantum channels limit the maximum distance to about fifty to hundred kilometers. The two main reasons for this limited reach are the inevitable depolarization of the photons and the intrinsic loss in optical fibers [GRTZ02]. Fortunately, theoretical proposals showed that this limitation may be overcome by means of *quantum repeaters* [BDCZ98, DLCZ01]. In these protocols, the quantum channel is divided into smaller sections of about the size of the coherence or absorption length of the optical fibers, and entanglement is propagated by a generalized quantum teleportation called *entanglement swapping* [ZZHE93]. Actually, the scheme is slightly more subtle since one cannot clone nor amplify the particles (*qubits*) carrying the quantum information without destroying their quantum nature [WZ82]. Although the implementation of large quantum repeaters remains an ambitious task, the essential elements needed to their realization are already established [PGU[+]03, YCZ[+]08], converting the magical thought of a perfectly secure communication between distant parties into an accessible technology.

In this Thesis, we study the distribution of entanglement in quantum networks in general, analyzing the generation of long-distance entanglement from an arrangement of short quantum connections in particular. We consider that neighboring stations share initially some *partially* entangled pairs of qubits, which emphasizes the difficulty of creating remote entanglement, and the task is then to design local quantum operations such that the entanglement present in the whole network gets concentrated between a small number of chosen stations. We will see that the geometry of the network plays a crucial role in this problem, but the real *leitmotiv* in this Thesis is that the best results are obtained if one "thinks quantumly" not only at the connection scale, but also from a global network perspective. In this respect, finding powerful entanglement-distribution strategies deeply depends on our knowledge of *multipartite entanglement* properties, that is, the characteristics of entangled states of more than two particles.

The first part of the Thesis deals with a pure-state description of the quantum networks.

In Chap. 1, we start by analyzing carefully the basic quantum operations that allow one to propagate or concentrate the entanglement present in the links of the network. We review the entanglement-swapping procedure and introduce some figures of merit to quantify its efficiency (Sec. 1.1). Then, we characterize the quantum measurements that optimally achieve this task, which leads to an interesting statement: already for systems in which only two entanglement swappings are performed, projecting onto partially-entangled states may yield better results than considering a basis of maximally-entangled pairs of qubits, see Sec. 1.2. Finally, we justify the model of the quantum connections by showing that the optimum success probability of entanglement generation between two distant stations in one-dimensional system decays exponentially with their distance (Sec. 1.3). In fact, this corresponds to the existence of an absorption (or depolarization) length in real optical fibers.

In Chap. 2, we turn to two-dimensional networks, in particular to lattices in which neighboring nodes share some pure-state entanglement. The goal is to generate an entangled pair of qubits between two distant stations, and the presence of many different paths between two nodes of the lattice greatly helps in this task. In fact, if the entanglement of the connections is larger than some critical value depending on the network structure, then it can be propagated over an arbitrarily large distance. We generalize some previously known results about *entanglement percolation*, showing that, in some cases, quantum measurements on the nodes can increase the efficiency of the strategy (Sec. 2.2). The idea of preprocessing the entanglement percolation by a judicious choice of local quantum operations, namely a projection onto multipartite entangled states, is further developed in Sec. 2.3. This leads to a systematic improvement of the creation of entanglement over a large distance, lowering the entanglement threshold regardless of the lattice geometry.

Introducing the concept of quantum complex networks (Chap. 3), we temporarily leave the problem of long-distance entanglement generation in lattices to focus on networks of richer topology. Three main properties of real-world communication networks, that are absent in regular lattices, are a small-world, clustering and scale-free behavior. The first mathematical model describing networks with the small-world property is the random graphs of Erdős and Rényi (Sec. 3.1.1), and in Sec. 3.1.2 we propose a natural extension to the quantum domain. The following sections aim to show that, mainly due to the superposition principle, the quantum complex networks behave completely distinct from their classical cousins.

The pure-states formalism brings a deep insight into the broad range of possibilities, but also the restrictions, of entanglement manipulation in quantum networks, However, realistic settings are best described by mixed states, since considering such quantum mixtures expresses our lack of control over all degrees of freedom of the system. The aim of Chap. 4 is twofold. First, it makes the transition between the two parts of this Thesis by showing how the results collected so far still apply for certain kinds of errors, or noise, that perturb any experiment (Sec. 4.1). Second, it briefly reviews the quantum-repeater protocols, which are proposed to create remote entanglement in noisy settings, drawing attention to their limitations (Sec. 4.2.1). In particular, the need of reliable quantum memories to store the qubits for a long time is one of the most severe problems they have to face.

A new scheme for generating long-distance entanglement in quantum networks subject to general noise is presented in Chap. 5. Making full use of the geometry of two-dimensional square lattices, we combine a classical and a topological error-correcting codes into one efficient quantum protocol. All operations are performed fault-tolerantly and, equally important, simultaneously (Sec. 5.2). Therefore, the qubits have to be preserved from decoherence for a short time only, which relaxes the requirement of good quantum memories. Moreover, the overhead of local resources increases very slowly with the distance, while the tolerable error probability stays on the order of one percent for any realistic network size, making our proposal favorable for quantum communication.

Finally, in Chap. 6, we prove that entanglement can be established between two infinitely distant qubits of a three-dimensional network if the noise of the connections is not too strong. To that end, we first transform the quantum state that describes the initial network into a *cluster state*, which is a highly-entangled multipartite state (Sec. 6.1). Then, it is shown how the information gained by measuring the qubits at each station allows one to correct most errors affecting the links of the network (Sec. 6.2). Since only a constant overhead of qubits is required per station, this strategy further lessens the physical resources that are needed for long-distance communication in quantum networks.

The various results presented in Chaps. 1 and 2 have been published in [PCA+08], except those of Sec. 2.3 which are to be found in [PCL+10]. The results of Chaps. 3, 5, and 6 and the related discussions in Chap. 4 have been published in [PLAC10], [PJS+08], and [Per10], respectively. This work has been supported by the Ph.D. program "Quantum Computing, Control and Communication" of the Elite Network of Bavaria.

PART I

Pure states

> *Making the simple complicated is commonplace. Making the complicated simple, awesomely simple, that's creativity.*
>
> — CHARLES MINGUS

CHAPTER 1

Entanglement manipulation in basic networks

The first part of this Thesis is devoted to quantum networks in which connections are described by two-qubit entangled pure states of the form

$$|\varphi\rangle \equiv \sqrt{\varphi_0}\,|00\rangle + \sqrt{\varphi_1}\,|11\rangle, \tag{1.1}$$

with $\varphi_0 + \varphi_1 = 1$ and where our convention is to choose $\varphi_0 \geq \varphi_1$. Setting this last inequality to be strict reflects the fact that remote entanglement cannot be generated perfectly in realistic settings. On the contrary, we consider that all stations have a complete and perfect control over their particles, so that no restriction is put on the quantum operations they can perform locally. For instance, we permit the use of ancillary qubits, and no error occurs while manipulating, storing, or measuring the quantum system at a station. This may seem to be a crude approximation, but it allows one to get a deep insight into the way entanglement can be manipulated in general quantum systems. In this chapter, we study the basics of entanglement manipulation in communication networks in the very spirit of [BVK98, HS00]. The aim is to investigate and derive optimal local measurement protocols for networks consisting of few qubits only. The results obtained for these simple situations will then be used as building blocks for more elaborated schemes in larger networks.

In Sec. 1.1, we start by describing the operation that propagates entanglement over a larger distance in a quantum network, namely the *entanglement swapping* [ZZHE93] at a node that is referred to as an *entanglement swapper* (or a *quantum relay*, see [dRMT+04]). This quantum operation involves a joint measurement on two qubits and is optimally performed in a basis of maximally entangled states. Depending on the figure of merit quantifying the efficiency of this procedure, however, we show that different bases (local rotation of the qubits) are to be used.

In Sec. 1.2, we consider quantum systems that consist of two entanglement swappers, and we find that projecting onto maximally-entangled states does not maximize, in general, the various figures of merit. In fact, the measurements at the middle stations are best performed in a basis of partially entangled states. This result, besides being rather surprising, will have some importance in the entanglement distribution protocols of the next chapter (Sec. 2.3.2). Then, we describe another useful quantum operation to manipulate the entanglement of pairs of qubits: while the entanglement swapping transfers (but inevitably decreases) the entanglement from one node to another, the *distillation* of several entangled states can concentrate it into one pair only. These two antagonistic effects will find a direct application in the deterministic creation of long-distance entanglement in quantum networks, see Sec. 2.1.

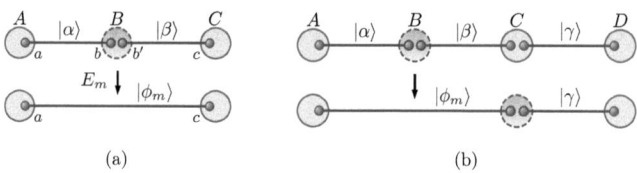

Figure 1.1: Examples of the *entanglement-swapping* procedure in quantum networks. (a) Entanglement can be generated between two previously unconnected stations A and C by an entanglement swapping at the middle station. We call the station B an *entanglement swapper* or a *quantum relay*. (b) Same consideration, but with two intermediate nodes.

Finally, in Sec. 1.3, we consider arbitrarily large chains of entanglement swappers. The probability to successfully generate an entangled pair of qubits between the extremities of the chain decreases exponentially with its length, which is a well-known result that is easily derived in our formalism. We then determine the exact value of the average entanglement for a specific choice of the measurement bases; it will be shown in Sec. 2.3.1 how this formula is related to the probability of creating some multipartite entangled states.

1.1 Entanglement swapping

In this section, we show how the entanglement present in single connections can be propagated over a larger distance. We introduce some figures of merit to quantify the efficiency of the procedure and describe the quantum operations yielding the optimum results. These constructions will be then extensively used to design powerful protocols for much larger systems.

1.1.1 Joint measurement at the middle station

It is a well-known result that any two-qubit entangled pure state can be transformed into the state described in Eq. (1.1) by performing a local basis rotation, which defines its Schmidt decomposition with coefficients φ_0 and φ_1. We refer the reader to [NC00], for example, for the elementary definitions and notions related to quantum information theory, so that only the strictly necessary notation has to be introduced here. In Fig. 1.1a, we depict one of the most primitive networks, but nevertheless important, that can be imagined. In this configuration, the central station applies a joint measurement on its two qubits, so that the extremities A and C become entangled. For instance, the station B can perform a measurement described by n positive operators E_m satisfying the completeness relation $\sum_{m=1}^n E_m = \mathbb{1}_4$, in which case the resulting (non-normalized) state of the outcome m reads

$$\rho_m \equiv \mathrm{tr}_{bb'}\left((\mathbb{1}_2 \otimes E_m \otimes \mathbb{1}_2)\, |\alpha\beta\rangle\langle\alpha\beta| \right)$$

and appears with probability $p_m = \text{tr}(\rho_m)$. Restricting to projective measurements only, that is, setting $E_m = |\mu_m\rangle\langle\mu_m|$ for some normalized state $|\mu_m\rangle = \sum_{i,j=0}^1 \mu_{m,ij}|ij\rangle$, one verifies that the smallest Schmidt coefficient of ρ_m is

$$\lambda_m = \frac{1}{p_m}\min\{\text{eigval}(\rho'_m)\} = \frac{1}{2}\left(1 - \sqrt{1 - \frac{4\det(\rho'_m)}{p_m^2}}\right), \quad (1.2)$$

with $\rho'_m \equiv \text{tr}_c(\rho_m)$. This formula can be written in a more compact way by considering the following map from the space of two-qubit pure states to that of 2×2 complex matrices:

$$|\nu\rangle = \sum_{i,j=0}^1 \nu_{ij}|ij\rangle \quad\mapsto\quad \hat{\nu} = \sum_{i,j=0}^1 \nu_{ij}|i\rangle\langle j|. \quad (1.3)$$

In fact, ρ'_m is now given by $M_m M_m^\dagger$ with $M_m = \hat{\alpha}\,\hat{\mu}_m^*\,\hat{\beta}$, and the quantities of interest read:

$$C_m = \frac{2|\det(M_m)|}{p_m} = \frac{\sqrt{\alpha_0\alpha_1\beta_0\beta_1}}{p_m}C(\mu_m), \quad (1.4\text{a})$$

$$\lambda_m = \frac{1}{2}\left(1 - \sqrt{1 - C_m^2}\right), \quad (1.4\text{b})$$

$$p_m = \sum_{i,j=0}^1 \alpha_i \beta_j |\mu_{m,ij}|^2, \quad (1.4\text{c})$$

where we have introduced the concurrence, a measure of entanglement for general states of two qubits [Woo98]. For pure states, the concurrence is defined as $C(\alpha) \equiv 2|\det(\hat{\alpha})|$. In this setting, the prototype of projective measurements is clearly the entanglement swapping, which teleports the qubit b to c by consuming the connection $|\beta\rangle$. The measurement of the qubits b and b' is performed in a Bell basis, which we define as follows. Starting from the computational basis $\{|0\rangle, |1\rangle\}$, we define two new orthogonal bases $\{|\uparrow\rangle, |\downarrow\rangle\}$ and $\{|\Phi^+\rangle, |\Phi^-\rangle, |\Psi^+\rangle, |\Psi^-\rangle\}$ for one and two qubits, respectively:

$$\begin{pmatrix}|\uparrow\rangle \\ |\downarrow\rangle\end{pmatrix} = U\begin{pmatrix}|0\rangle \\ |1\rangle\end{pmatrix}, \quad (1.5\text{a})$$

with $U \in \mathcal{U}(2)$, and

$$|\Phi^\pm\rangle \equiv \frac{|\uparrow\uparrow\rangle \pm |\downarrow\downarrow\rangle}{\sqrt{2}}, \quad |\Psi^\pm\rangle \equiv \frac{|\uparrow\downarrow\rangle \pm |\downarrow\uparrow\rangle}{\sqrt{2}}. \quad (1.5\text{b})$$

The latter four vectors are known as the *Bell states* if no local rotation is performed, *i.e.*, if $U = \mathbb{1}_2$ for both qubits. In this case, $|\uparrow\rangle$ and $|\downarrow\rangle$ are the eigenvectors of the Pauli matrix Z, and we call the basis indistinctly the Bell or the ZZ basis. Another two-qubit basis plays a key role while manipulating entangled states: the XZ basis, in which the first unitary corresponds to the Hadamard matrix, so that $|\uparrow\rangle$ and $|\downarrow\rangle$ are the eigenvectors of the Pauli matrix X for the first qubit. Explicitly, the ZZ and XZ bases are given by the columns of the matrices

$$M_{\text{ZZ}} = \frac{1}{\sqrt{2}} \begin{pmatrix} 1 & 1 & 0 & 0 \\ 0 & 0 & 1 & 1 \\ 0 & 0 & 1 & -1 \\ 1 & -1 & 0 & 0 \end{pmatrix} \text{ and } M_{\text{XZ}} = \frac{1}{2} \begin{pmatrix} 1 & 1 & 1 & -1 \\ 1 & -1 & 1 & 1 \\ 1 & 1 & -1 & 1 \\ -1 & 1 & 1 & 1 \end{pmatrix}. \tag{1.6}$$

1.1.2 Figures of merit

We describe now three figures of merit that are used to evaluate the usefulness of an entanglement distribution protocol: the singlet conversion probability (SCP), the worst-case entanglement (WCE), and the average concurrence. These figures of merit take value in the interval $[0, 1]$ and are somehow related to each other, but they present some subtle differences nonetheless. Because they are intimately related to the transformation of bipartite pure states under local operations and classical communication (LOCC), we first recall the connection between the Schmidt coefficients and majorization theory, which will naturally lead to the definition of our figures of merit.

LOCC transformations and majorization theory

Consider two pure states $|\alpha\rangle$ and $|\beta\rangle$ in a bipartite system. Can $|\alpha\rangle$ be transformed into $|\beta\rangle$ by LOCC in a deterministic way? The solution to this question was found in 1999 by Nielsen [Nie99], who noticed a connection between this problem and majorization theory. Based on this relation, Vidal extended the results to obtain the optimal probability for LOCC conversion between states whenever a deterministic transformation is impossible [Vid99]. We briefly review these results here since this formalism is widely employed in many of the protocols described in the next sections. The interesting reader is referred to [NV01] for more details.

Let us introduce the concept of majorization by considering two d-dimensional real vectors, $\boldsymbol{v} = (v_0, \ldots, v_{d-1})$ and $\boldsymbol{w} = (w_0, \ldots, w_{d-1})$, whose components are positive, sum up to one, and are sorted in decreasing order. Then $\boldsymbol{v}$ is said to be majorized by $\boldsymbol{w}$, which is denoted by $\boldsymbol{v} \prec \boldsymbol{w}$, if the inequalities

$$\sum_{i=0}^{l} v_i \leq \sum_{i=0}^{l} w_i \tag{1.7}$$

hold for all $l \in \{0, \ldots, d-1\}$. The beautiful connection to entanglement manipulation between bipartite pure states is that $|\alpha\rangle$ can be transformed into $|\beta\rangle$ with unit probability whenever $\boldsymbol{\alpha} \prec \boldsymbol{\beta}$, where $\boldsymbol{\alpha}$ is the vector of Schmidt coefficients of $|\alpha\rangle$ (and similarly for $\boldsymbol{\beta}$). Moving now to non-deterministic transformations, the optimal probability for LOCC conversion is

$$p(\alpha \to \beta) = \min_{l} \left\{ \frac{\sum_{i=0}^{l} \alpha_i}{\sum_{i=0}^{l} \beta_i} \right\}. \tag{1.8}$$

SCP, WCE and concurrence

A direct application of Eq. (1.8) is that a two-qubit state $|\alpha\rangle$, as defined in Eq. (1.1), can be converted into a singlet, or more generally into a Bell pair, with optimal probability

$$S(\alpha) \equiv \text{prob}\left(|\alpha\rangle \mapsto |\Psi^-\rangle\right) = 2\alpha_1. \tag{1.9}$$

Explicitly, this result (also known as the "Procrustean method" of entanglement concentration [BBPS96]) is obtained by performing on one of the qubits a generalized measurement defined by the operators

$$M_1 = \begin{pmatrix} \sqrt{\frac{\alpha_1}{\alpha_0}} & 0 \\ 0 & 1 \end{pmatrix} \quad \text{and} \quad M_2 = \begin{pmatrix} \sqrt{1 - \frac{\alpha_1}{\alpha_0}} & 0 \\ 0 & 0 \end{pmatrix}. \tag{1.10}$$

Returning to the one-swapper configuration in which the central station applies some measurement operators E_m, we define the corresponding average SCP as

$$S^{(2)}(\alpha, \beta) \equiv \sum_{m=1}^{n} p_m S(\phi_m) = 2 \sum_{m=1}^{n} p_m \phi_{m,1}, \tag{1.11}$$

where p_m is the probability that $|\phi_m\rangle$ is generated. This equation can be generalized to a system of $N-1$ entanglement swappers, in which case the figure of merit is denoted by $S^{(N)}$; if the context is clear, however, we simply write S. One can also be interested in maximizing the entanglement for all outcomes, which is characterized by the WCE:

$$W \equiv \min_m \{S(\phi_m)\}. \tag{1.12}$$

Finally, in the same spirit, we define the average concurrence of the measurement to be

$$C \equiv \sum_{m=1}^{n} p_m C_m. \tag{1.13}$$

In these equations, the figures of merit are defined for one measurement only, but the generalization to larger systems is straightforward: one has just to let the sum run over all possible outcomes of the various measurements.

1.1.3 Optimal measurements

The power of the protocols described in the next chapters depends crucially on the efficiency of the entanglement swappings, and the choice of the measurement basis is thus of special interest. In that respect, although the Bell basis could be parameterized by four angles describing two one-qubit rotations, it turns out that calculations are much easier when done in the "magic basis" [HW97]

$$(\,|\Phi_1\rangle,\,|\Phi_2\rangle,\,|\Phi_3\rangle,\,|\Phi_4\rangle) \equiv (\,|\Phi^+\rangle, -i\,|\Phi^-\rangle, -i\,|\Psi^+\rangle,\,|\Psi^-\rangle). \tag{1.14}$$

In this basis, in fact, the concurrence of a state $|\nu\rangle = \sum_{i=1}^{4} \nu_i\,|\Phi_i\rangle$ simply reads $C(\nu) = \left|\sum_{i=1}^{4} \nu_i^2\right|$. Since the concurrence of a Bell state is one, it follows that its components in the magic basis have all the same phase. Thus, they can be restricted to take value within the set of the real numbers. Let $\{|\mu_m\rangle\}$ be a set of four orthogonal such states. Then the outcome probabilities given in Eq. (1.4c) read

$$p_m = p_{\min}\left(\mu_{m,1}^2 + \mu_{m,2}^2\right) + p_{\max}\left(\mu_{m,3}^2 + \mu_{m,4}^2\right), \tag{1.15a}$$

with

$$p_{\min} \equiv \frac{\alpha_0\beta_1 + \alpha_1\beta_0}{2} \quad \text{and} \quad p_{\max} \equiv \frac{\alpha_0\beta_0 + \alpha_1\beta_1}{2}. \tag{1.15b}$$

It has to be emphasized that, given $|\alpha\rangle$ and $|\beta\rangle$ and for $C(\mu_m) = 1$, these probabilities fully characterize a Bell measurement since, from Eq. (1.4), the Schmidt coefficients of the resulting states depend on p_m only:

$$\lambda_m = \frac{1}{2}\left(1 - \sqrt{1 - \frac{\alpha_0\alpha_1\beta_0\beta_1}{p_m^2}}\right). \tag{1.16}$$

Remark that any measurement in the Bell basis yields four values p_m that lie in the interval $[p_{\min}, p_{\max}]$, and let us now state and prove a reverse proposition that turns out to be very useful while maximizing the SCP:

Proposition 1.1 (Bell measurements and outcome probabilities)

Let $\{x_m\}$ be a set of four real numbers that add up to one and that lie in the interval $[p_{\min}, p_{\max}]$. Then there exists a Bell measurement whose outcome probabilities p_m are equal to x_m.

Proof First, remark that the conditions on $\{x_m\}$ are clearly necessary since it describes a probability distribution, and because $p_m = p_{\min}\kappa_m + p_{\max}(1 - \kappa_m)$ for some $\kappa_m \in [0,1]$, see Eq. (1.15a). The degenerate situation $p_{\min} = p_{\max} = 1/4$ is trivial since it only arises if both $|\alpha\rangle$ and $|\beta\rangle$ are maximally entangled. In this case, all Bell measurements yield the same outcome probabilities $p_m = 1/4$. We thus consider $p_{\min} < p_{\max}$, and without loss of generality we sort the x_m by increasing value: $x_i \leq x_{i+1}$. Let us now denote by $\{\mu_m\}$ the four vectors describing the Bell measurement in the magic basis. Since they are orthogonal, one of them is completely determined (up to a sign) by the three others; let μ_4 be this vector. We parametrize the first three vectors as

$$\mu_m = \left(\sqrt{\kappa_m}\cos(\theta_m), \sqrt{\kappa_m}\sin(\theta_m), \sqrt{1-\kappa_m}\cos(\omega_m), \sqrt{1-\kappa_m}\sin(\omega_m)\right),$$

with $\kappa_m = (p_{\max} - x_m)/(p_{\max} - p_{\min})$, so that they are normalized and satisfy $p_m = x_m$. The remaining task is to prove that there always exist some angles θ_m and ω_m leading to an orthogonal basis. This is straightforward if $\kappa_m \in \{0, 1\}$ for some m, but let us write the orthogonality conditions as

$$\begin{cases} \cos(\theta_1 - \theta_2) = -\eta_1\eta_2 \cos(\omega_1 - \omega_2), \\ \cos(\theta_1 - \theta_3) = -\eta_1\eta_3 \cos(\omega_1 - \omega_3), \\ \cos(\theta_2 - \theta_3) = -\eta_2\eta_3 \cos(\omega_2 - \omega_3), \end{cases} \quad (1.17)$$

with $\eta_m = \sqrt{(1-\kappa_m)/\kappa_m}$. In this system, only four angles are relevant: $\theta_a = \theta_1 - \theta_2$, $\theta_b = \theta_1 - \theta_3$, $\omega_a = \omega_1 - \omega_2$ and $\omega_b = \omega_1 - \omega_3$, which can be freely chosen in the interval $[0, \pi]$. There are, however, some constraints on these variables. In fact, $\kappa_i \geq \kappa_{i+1}$ and $\sum_i \kappa_i = 2$, so that $\eta_1\eta_2 \leq 1$ and $\eta_1\eta_3 \leq 1$. Therefore we have $\theta_{a,b} \in [\theta^*_{a,b}, \pi - \theta^*_{a,b}]$, with $\theta^*_{a,b} \in [0, \frac{\pi}{2}]$ such that $\cos(\theta^*_a) = \eta_1\eta_2$ and $\cos(\theta^*_b) = \eta_1\eta_3$. It follows that $\cos(\theta_2 - \theta_3)$ is limited to the range $[-\cos(\theta^*_a + \theta^*_b), 1]$, and one can check that there always exists at least one solution to the orthogonality conditions except if $\cos(\theta^*_a + \theta^*_b) < -\eta_2\eta_3$, which however never happens. In fact, this last inequality can be written in terms of κ_1, κ_2 and κ_3 only, which leads, after some tedious algebra, to $\kappa_1 + \kappa_2 + \kappa_3 > 2$. But the sum of the κ's equals two, and thus there always exists an orthonormal basis of Bell vectors leading to the probability distribution $\{x_m\}$. □

We have now all the tools to find the quantum operations that maximize the figures of merit of an entanglement swapping. Let us state the results in the following three propositions:

Proposition 1.2 (Measurement optimizing the SCP)

The measurement that maximizes the SCP after one entanglement swapping is the Bell measurement in the ZZ basis, and

$$S_{\max} = 2\min\{\alpha_1, \beta_1\}. \quad (1.18)$$

Proof Two kinds of outcomes appear when performing a Bell measurement in the ZZ basis: two of the outcome probabilities are equal to $p_{\max}$, while the other two are equal to $p_{\min}$. Inserting these values into Eq. (1.4), one finds the corresponding smallest Schmidt coefficients:

$$\lambda(p_{\min}) = \frac{\min\{\alpha_0\beta_1, \alpha_1\beta_0\}}{2 p_{\min}}, \quad \lambda(p_{\max}) = \frac{\alpha_1\beta_1}{2 p_{\max}}, \quad (1.19)$$

whence $S_{ZZ} = 2\min\{\alpha_1, \beta_1\}$. Consider now that we are allowed to jointly perform an arbitrary operation not only on b and b', but also on a. We are in the presence of a bipartite system, such that the results of majorization theory apply: the SCP of this system is at most $2\beta_1$. A similar construction for the qubits b, b' and c tells us that the SCP is at most $2\alpha_1$, so that the final SCP cannot exceed twice the minimum of α_1 and β_1. □

Remark that the SCP does not decrease after one entanglement swapping if we set $\alpha = \beta$. This is the "conserved entanglement" described in [BVK99].

Proposition 1.3 (Measurement optimizing the WCE)

The measurement that maximizes the WCE for a one-swapper configuration is the Bell measurement in the XZ basis:

$$W_{\max} = 1 - \sqrt{1 - 16\,\alpha_0\alpha_1\beta_0\beta_1}. \tag{1.20}$$

Proof Supposing that the best measurement is done in the Bell basis, the result directly follows from Prop. 1.1 and Eqs. (1.6) and (1.16). In fact, one has to find a measurement that leads to four identical outcome probabilities $p_m = 1/4$, which is achieved in the XZ basis. Now, let us prove the proposition by contradiction. Suppose that there exists a measurement described by the operators $\{E_m = |u_m\rangle\langle u_m|\}_{m=1}^n$, with $n \geq 4$, such that $W_E > W_{XZ}$. Then each λ_m has to be strictly greater than the smallest Schmidt coefficient of the outcomes in the XZ basis. Thus, from Eq. (1.2),

$$\det(\tilde{\rho}_m) > p_m^2\, 4\,\alpha_0\alpha_1\beta_0\beta_1 \quad \forall\, m. \tag{1.21}$$

Since $\det(\tilde{\rho}_m) = \alpha_0\alpha_1\beta_0\beta_1 \,|\det(\hat{u}_m)|^2$, summing over m the square root of this last equation yields

$$\sum_{m=1}^n |\det(\hat{u}_m)| > 2. \tag{1.22}$$

We know that the concurrence of a (normalized) state is smaller than or equal to one, hence $2\,|\det(\hat{u}_m)| \leq \|u_m\|^2$. Taking the trace of the completeness relation for the operators E_m further implies $\sum_{m=1}^n \|u_m\|^2 = 4$. Therefore, we have $\sum_{m=1}^n |\det(\hat{u}_m)| \leq 2$, which is in contradiction with Eq. (1.22) and thus concludes the proof. $\square$

Proposition 1.4 (Measurement optimizing the concurrence)

Any Bell measurement maximizes the average concurrence of one entanglement swapping:

$$C_{\max} = 2\sqrt{\alpha_0\alpha_1\beta_0\beta_1}. \tag{1.23}$$

Proof It is clear from Eqs. (1.4a) and (1.13) that the maximum average concurrence is obtained for $C(\mu_m) = 1$ for all outcomes, which describes, by definition, a Bell measurement in any basis. $\square$

1.2 Bases of maximally-entangled states are not always optimum

Entanglement is known to lie at the root of quantum communication and is often argued to be at the origin of any speed-up for quantum computation. Therefore, it seems that maximally-

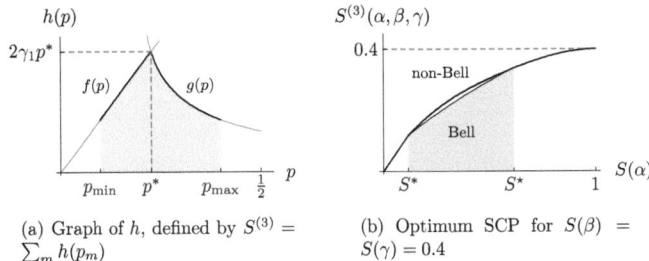

Figure 1.2: SCP after two consecutive entanglement swappings. (a) Function $h(p) = \min\{f(p), g(p)\}$, and definition of p^*. (b) Optimal SCP as a function of the entanglement of $|\alpha\rangle$. Numerical results show that there exists a better strategy than the Bell measurements (thin line) for $S(\alpha) \in \,]S^*, S^\star[$. The values S^* and $S^\star$ are such that $p^*(\alpha^*) = p_{\min}(\alpha^*)$ and $p^*(\alpha^\star) = (1 - p_{\max}(\alpha^\star))/3$.

entangled states should be, in some sense, always the best ones to perform any basic quantum task. We have proven that this is indeed the case for one entanglement swapping, where bases of Bell pairs are used to optimize the various figures of merit. In this section, however, we show that this statement is not true in general. To that end, we investigate some systems that consist of two entanglement swappings only. An analogous result has been reported recently by Modławska and Grudka [MG08], who have demonstrated that non-maximally entangled states can be better for the realization of multiple linear optical teleportation in the scheme of Knill, Laflamme and Milburn [KLM01]. Similarly, in the context of universal quantum computation, the authors of [GFE09] conclude that entanglement must "come in the right dose".

1.2.1 Two consecutive entanglement swappings

We consider here a system of three collinear states on which we perform two consecutive measurements, as depicted in Fig. 1.1b. In what follows, we describe the measurements that maximize the figures of merit introduced in the previous section.

The maximization of the WCE and the concurrence is somehow trivial for a two-swapper configuration. First, any Bell measurement maximizes the average concurrence of the results of the two measurements. This will be generalized and proven in Sec. 1.3.1 for an arbitrary number of entanglement swappings. Second, in order to maximize the WCE, we simply perform a XZ measurement at the two central nodes. In fact, if one performs any other measurement on the first quantum relay, then at least one resulting state $|\phi_m\rangle$ is less entangled than the corresponding XZ results. This further reflects in the WCE of the second measurement, which has to be a Bell measurement in the XZ basis from Prop 1.3.

Let us now turn to the more interesting problem of maximizing $S^{(3)}$, the average SCP of a chain of three entangled pairs of qubits and two quantum relays. The first measurement yields some states $|\phi_m\rangle$ with outcome probabilities p_m, while from Prop. 1.2 we already know that the second measurement has to be done in the ZZ basis. Hence, we have to find for the first quantum relay the measurement operators E_m that maximize

$$S^{(3)}(\alpha, \beta, \gamma) \equiv 2 \sum_m p_m \min\{\varphi_{m,1}, \gamma_1\}. \qquad (1.24)$$

We first maximize this quantity over the set of Bell measurements[1] and then present some numerical results showing that non-Bell measurements sometimes yield higher values of the SCP.

Bell measurements

We fix the states α, β, and γ and consider the SCP as a function of the outcome probabilities only:

$$S^{(3)}(\{p_m\}) = \sum_m \min\{f(p_m), g(p_m)\} \equiv \sum_m h(p_m), \qquad (1.25)$$

where $f(p) \equiv 2\gamma_1 p$ and $g(p) \equiv p - \sqrt{p^2 - \alpha_0\alpha_1\beta_0\beta_1}$. One can easily verify that $g(p)$ is strictly decreasing and convex for any states α and β:

$$\frac{d}{dp}g(p) < 0 \quad \text{and} \quad \frac{d^2}{dp^2}g(p) > 0 \quad \forall p \in [p_{\min}, p_{\max}].$$

A typical example of $h(p)$ is plotted in Fig. 1.2a, and the value p^* at which the two functions f and g cross each other is given by:

$$p^* = \frac{1}{2}\sqrt{\frac{\alpha_0\alpha_1\beta_0\beta_1}{\gamma_0\gamma_1}}. \qquad (1.26)$$

It is sufficient to maximize the function $S^{(3)}$ over the set of all possible probability distributions, since Prop. 1.1 ensures the existence of a Bell measurement leading to any such distribution. At this point, let us give three conditions that have to be satisfied by the best probability distribution:

(i) Obviously, $p_m \in [p_{\min}, p_{\max}]$ for all outcomes m, and $\sum_m p_m = 1$.

(ii) If the set $\{p_m\}$ maximizes the SCP, then all probabilities lie either to the left or to the right of p^*. In fact, suppose for example that $p_1 + 2\varepsilon < p^* < p_2 - 2\varepsilon$ and choose $\tilde{p}_1 = p_1 + \varepsilon$ and $\tilde{p}_2 = p_2 - \varepsilon$, with $0 < \varepsilon \ll 1$. The constraints on these new probabilities are still satisfied, but a better SCP is found.

(iii) Since g is convex, if p_1 and p_2 are such that $p^* + 2\varepsilon < p_1 \leq p_2 < p_{\max} - 2\varepsilon$, then the choice $\tilde{p}_1 = p_1 - \varepsilon$ and $\tilde{p}_2 = p_2 + \varepsilon$ yields a strictly greater SCP.

---

[1] As we will see, this is the best strategy for a wide range of entanglement in the connections α, β and γ.

p^* versus $p_{\min}$ and $p_{\max}$	$\{p_m\}$ maximizing $S^{(3)}$	strategy
$p^* \leq p_{\min}$	$\{p_{\min}, p_{\min}, p_{\max}, p_{\max}\}$	ZZ
$p_{\min} \leq p^* \leq \frac{1}{3}(1 - p_{\max})$	$\{p^*, p^*, p_{\max}, 1 - 2p^* - p_{\max}\}$	non-Bell
$\frac{1}{3}(1 - p_{\max}) \leq p^* \leq \frac{1}{4}$	$\{p^*, p^*, p^*, 1 - 3p^*\}$	Bell
$p^* \geq \frac{1}{4}$	$\{\frac{1}{4}, \frac{1}{4}, \frac{1}{4}, \frac{1}{4}\}$	XZ

Table 1.1: Maximization of $S^{(3)}$ over the set of Bell measurements (2nd column), or performing arbitrary measurements (3rd column).

From these considerations, it is now very simple to maximize $S^{(3)}$, and one sees that the value p^*, with respect to $p_{\min}$ and $p_{\max}$, plays a crucial role in the choice of the optimal measurement. In fact, we have to distinguish four cases, as described in Tab. 1.1. Note that ZZ measurements lead to the maximum SCP whenever $p^* \leq p_{\min}$, while the XZ basis is the best strategy when $p^* \geq 1/4$. So far, we have maximized the SCP after two entanglement swappings supposing that the first measurement had to be done on the states α and β. But what happens if we start from the right extremity of the chain? It appears that the maximum SCP depends, in general, on the order of the measurements, and that performing the first measurement on the most entangled states yields better results.

General measurements

The question that has still to be answered is whether or not considering bases of non-maximally entangled states leads to a better SCP than the Bell measurements. Since the concurrence of the states used for the entanglement swapping takes now any value between 0 and 1, we cannot consider $S^{(3)}$ as a function of the outcome probabilities only. But for a fixed concurrence $C < 1$ we have:

$$\bar{g}(C, p) \equiv p - \sqrt{p^2 - \alpha_0 \alpha_1 \beta_0 \beta_1 \, C^2} < g(p) \quad \forall p.$$

Marking with a bar all variables of the general measurements, we have $\bar{p}^* < p^*$ and $\bar{g}(C, \bar{p}^*) < g(p^*)$. Therefore, one checks that the Bell measurements are indeed optimal except when $p_{\min} \leq p^* \leq (1 - p_{\max})/3$. The key fact about the Bell measurements in this case is that the three outcome probabilities cannot be chosen to lie on p^*, since the fourth one would be greater than $p_{\max}$. But the range of possible outcome probabilities depends on the concurrence: for example, from Eq. (1.4c), we have that $\bar{p}_m \in [\alpha_1 \beta_1, \alpha_0 \beta_0]$ for $C(u_m) = 0$, or more generally:

$$\bar{p}_m \in [\bar{p}_{\max}, \bar{p}_{\min}] \supseteq [p_{\max}, p_{\min}]. \tag{1.27}$$

Hence, a better strategy is to perform a measurement such that three outcomes probabilities are equal to $\bar{p}^*$ and such that the concurrences of the states are the largest ones satisfying

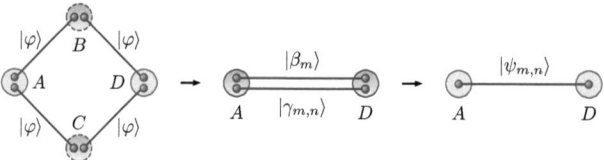

Figure 1.3: Operations on a single square network to obtain an entangled pair on its diagonal: measurements are performed at the two middle nodes, and the resulting states are distilled.

$\bar{p}_{\max} = 1 - 3\bar{p}^*$. This is confirmed by a numerical optimization of the first measurement, see Fig. 1.2b, and therefore it is sometimes advantageous to use a basis of non-maximally entangled pairs of qubits to perform the entanglement swappings.

1.2.2 A single square network

We study in this section a square whose borders are four identically entangled states, see Fig. 1.3. This is clearly one of the simplest possible two-dimensional networks, and thus it is a first step towards larger systems. The task is to entangle the two opposite nodes A and D, which is done in three steps. First, the station B performs a measurement on its two qubits, and a state $|\beta_m\rangle$ is created on the diagonal. Second, the station C measures its qubits in a basis that depends on m, gets a random outcome n, and therefore connects A and D with another state $|\gamma_{m,n}\rangle$. Third, the entanglement of these two states is concentrated into one two-qubit system, a procedure which is called *distillation*, yielding the state $|\psi_{m,n}\rangle$. The goal is of course to optimize the average entanglement of the final state, given the initial entangled links $|\varphi\rangle$. To that end, let us proceed backwards, starting with the optimization of the distillation.

Distillation

We temporarily fix the outcomes m and n, and without loss of generality we ask the corresponding Schmidt coefficients to satisfy $\beta_0 \geq \gamma_0$ (all formulas that follow are symmetric). A straightforward application of majorization theory gives us the conditions to distill the maximum amount of entanglement from β and γ in a deterministic way:

$$(\beta_0\gamma_0, \beta_0\gamma_1, \beta_1\gamma_0, \beta_1\gamma_1) \prec (\psi_0, \psi_1, 0, 0), \tag{1.28}$$

and the only non-trivial inequality that arises from Eq. (1.7) is $\beta_0\gamma_0 \leq \psi_0$. Therefore, the greatest Schmidt coefficient of the final state reads

$$\psi_0 = \max\left\{\frac{1}{2}, \beta_0\gamma_0\right\}. \tag{1.29}$$

Optimal measurement at the station C

The arguments used for the maximization of the WCE and the concurrence of a two-swapper configuration still hold here, so that XZ measurements are optimum for these two figures of merit. Applying such a measurement, we notice that a maximally entangled pair is obtained with unit probability if $\beta_0 \leq (2\beta_0^\star)^{-1}$, where

$$\beta_0^\star \equiv \frac{1+\sqrt{1-(4\varphi_0\varphi_1)^2}}{2}. \tag{1.30}$$

Concerning the SCP, we consider a given outcome m at the station B and write the function to maximize as

$$S^\triangle(\varphi,\beta) \equiv 2\sum_n p_n \left(1 - \max\left\{\frac{1}{2}, \beta_0\gamma_{n,0}\right\}\right) = 2\beta_1 + \beta_0 2\sum_n p_n \min\left\{\gamma_{n,1}, \frac{\beta_0 - \beta_1}{2\beta_0}\right\}$$
$$= S(\beta) + \beta_0 S^{(3)}(\varphi,\varphi,\beta'), \tag{1.31}$$

with $\beta_0' \equiv (2\beta_0)^{-1}$. Consequently, all results of Sec. 1.2.1 apply here too, and the important quantities introduced in that section read $p_{\min} = \varphi_0\varphi_1$, $p_{\max} = \frac{1}{2} - p_{\min}$, and

$$p^* = \frac{\varphi_0\varphi_1}{2\sqrt{\beta_0'\beta_1'}} = p_{\min}\frac{\beta_0}{\sqrt{\beta_0 - \beta_1}}. \tag{1.32}$$

Since p^* is larger than $p_{\min}$ for all β and φ, it follows that $S^\triangle_{\max}$ is reached by Bell measurements except when $p^* < (1 - p_{\max})/3$, see Tab. 1.1.

Optimizing the first measurement

Following the best WCE strategy, one finds that the largest Schmidt coefficient of the final state is given by $\psi_0 = \max\left\{1/2, \beta_0^{\star 2}\right\}$. Therefore, a perfect Bell pair can be generated between A and D if $\beta_0^\star \leq 1/\sqrt{2}$, that is, if

$$\varphi_0 \leq \varphi_0^\star \equiv \frac{1+\sqrt{1-\sqrt{2(\sqrt{2}-1)}}}{2} \approx 0.6498. \tag{1.33}$$

If this last inequality does not hold, one has to find the measurement at B that maximizes the function

$$S^\square(\varphi) \equiv \sum_m p_m S^\triangle_{\max}(\varphi,\beta_m), \tag{1.34}$$

where β_m is related to φ via Eq. (1.4). Now, we proceed exactly as for the two-swapper configuration: on the one hand we maximize $S^\square$ over the set of Bell measurements, and on the other hand we show that there exist some measurement bases yielding better results for certain states $|\varphi\rangle$. In the case of Bell measurements, we write $S^\square = \sum_m h(p_m)$, exactly as in Eq. (1.25), and

Entanglement of $|\varphi\rangle$	$\{p_m\}$ maximizing $S^\square$	strategy
$\varphi_0 \geq \varphi_0^\star$	$\{p^*, p^*, p_{\max}, 1 - 2p^* - p_{\max}\}$	non-Bell
$\varphi_0^\star \leq \varphi_0 \leq \varphi_0^*$	$\{p^*, p^*, p^*, 1 - 3p^*\}$	Bell
$\varphi_0 \leq \varphi_0^\star$	$\{\tfrac{1}{4}, \tfrac{1}{4}, \tfrac{1}{4}, \tfrac{1}{4}\}$	XZ

Table 1.2: Bell measurements at the station B that maximize $S^\square$, depending on the amount of entanglement in the connections $|\varphi\rangle$.

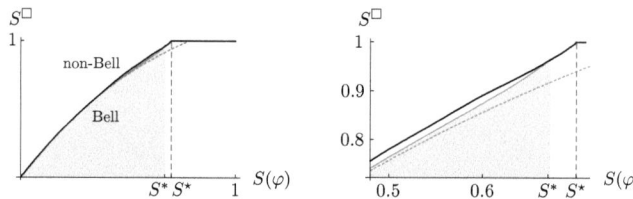

Figure 1.4: Maximum SCP of the single square network. Bell measurements (thin line delimiting the shaded area) are suboptimal for the whole range of entanglement $S(\varphi) < S^* \approx 0.672$, while a perfect singlet is obtained by performing XZ measurements if $S(\varphi) \geq S^* \approx 0.7$. Bell measurements in the ZZ basis (thin dashed line) are optimal in the regime of vanishing entanglement.

we optimize the function over the probabilities distributions $\{p_m\}$. Here we make a slight abuse of notation since we denote by the same character h two different functions, which however share many similar mathematical properties. In particular, the convexity arguments used in Sec. 1.2.1 still hold in the present case. The quantity p^* is now given by Eq. (1.32) by setting β to β^*, and therefore depends on φ only. In order to distinguish the various measurement strategies according to the amount of entanglement in the network, let us introduce the critical value $\varphi_0^\star \approx 0.664$ such that $p^* = (1 - p_{\max})/3$. With all these definitions, one finds the optimal measurements described in Tab. 1.2, and the similarity with Tab. 1.1 is immediate. Finally, numerical results attest that Bell measurements are suboptimal in the whole range $\varphi_0 > \varphi_0^*$, i.e., in the regime of weak entanglement. Remark, however, that measuring in the ZZ basis for nearly separable states $|\varphi\rangle$ approaches the optimal strategy, see Fig. 1.4.

1.3 An infinite chain of quantum relays

Let us conclude this first chapter by considering a chain of N entangled pairs of qubits attached to $N - 1$ nodes, which is the archetype of quantum communication networks since it connects arbitrarily distant stations in the most straightforward fashion. However, this one-dimensional system suffers an extremely severe limitation on the reachable communication distance: for partially entangled states, the probability of successfully generating a Bell pair between the two extremities of the chain decreases exponentially with N. Since this is one of the fundamental problems that are addressed in this Thesis, we carefully prove this statement in what follows. Moreover, the SCP after N entanglement swappings in the ZZ basis is explicitly calculated for all N, a result that will be used in Sec. 2.3.1 for more advanced strategies in two-dimensional networks.

1.3.1 Exponential decay of the entanglement

For simplicity, let us consider that all connections have the same amount of entanglement, such that the initial chain is given by the state $|\varphi\rangle^{\otimes N}$. A direct generalization [VMDC04] of Eqs. (1.4a) and (1.13) yields the following result for the average concurrence of the final state:

$$C^{(N)} = \sum_m 2 \, |\det(M_m)|, \tag{1.35}$$

with $M_m = \hat{\varphi}\, \hat{\mu}_{m_1}^* \, \hat{\varphi} \ldots \hat{\mu}_{m_{N-1}}^* \, \hat{\varphi}$, and where the states $|\mu_{m_i}\rangle$ are associated with the measurement result m_i of the i-th node. With these definitions, the maximization of $C^{(N)}$ over the measurements $\mathcal{M} \equiv \{|\mu_{m_i}\rangle\}$ reads:

$$\max_{\mathcal{M}} \left\{ C^{(N)} \right\} = 2\,|\det(\hat{\varphi})|^N \, \max_{\mathcal{M}} \left\{ \sum_m \left| \det(\hat{\mu}_{m_1} \ldots \hat{\mu}_{m_{N-1}}) \right| \right\} = |2\det(\hat{\varphi})|^N, \tag{1.36}$$

which holds for any bases of Bell pairs, i.e., if $|2\det(\hat{\mu}_{m_i})| = 1 \ \forall m_i$. Therefore, for non-maximally entangled connections $|\varphi\rangle$, the concurrence decreases exponentially with N:

$$C_{\max}^{(N)} = (4\,\varphi_0\varphi_1)^{N/2}. \tag{1.37}$$

Since the concurrence of a two-qubit pure state is always larger than or equal to twice its smallest Schmidt coefficient,[2] it follows that the other two figures of merit, namely the WCE and the SCP, also decay (at least) exponentially with N. For instance, a straightforward calculation shows that $W_{\max}$ behaves asymptotically as $(4\varphi_0\varphi_1)^N$ if one performs all measurements in the XZ Bell basis. This result is easily understood from the following observation: after many swappings, the entanglement of most outcomes becomes very weak, in which case the concurrence of the long-distance entangled pairs is proportional to the square root of their smallest Schmidt coefficient, leading to $W \sim C^2$. We do not have an exact formula for the maximum SCP (in the case of two entanglement swappings we already had to use numerics), but let us derive in the next section an explicit formula for the ZZ strategy.

1.3.2 Asymptotic behavior of the SCP under ZZ measurements

Measurements in the ZZ basis have been shown to maximize the SCP after one entanglement swapping (Prop. 1.2) and to be close to optimality for two quantum relays and for the single

---

[2] In fact, one has $C(\varphi) \equiv 2\sqrt{\varphi_0\varphi_1} \geq 2\varphi_1$ since $\varphi_0 \geq \varphi_1$.

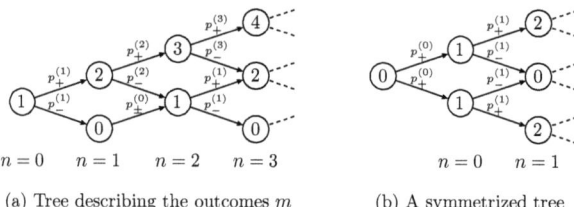

Figure 1.5: Resulting states $|\varphi^{(m)}\rangle$ after n consecutive entanglement swappings in the ZZ basis. (a) Weighted and directed tree, in which the nodes describe the labels m. (b) This tree can be made symmetric, with its root at $n = -1$, using the fact that $p_-^{(0)} = p_+^{(0)} = 1/2$.

square network in the regime of weak entanglement, see Tabs. 1.1 and 1.2. It is therefore natural to consider this startegy for an infinite chain of entangled pairs of qubits, and let us compute the average SCP of the system after N swappings. First, even if the number of outcomes grows exponentially with the length of the chain, one can keep track of all of them in an compact way. In fact, any state resulting from $n \leq N$ entanglement swappings in the ZZ basis has the form (up to local unitaries)

$$|\varphi^{(m)}\rangle \equiv \frac{\sqrt{\varphi_0^m}|00\rangle + \sqrt{\varphi_1^m}|11\rangle}{\sqrt{\varphi_0^m + \varphi_1^m}}, \quad m \in \mathbb{N}. \quad (1.38)$$

This is proven by induction on n, the case $n = 0$ corresponding to the initial state $m = 1$. Suppose that the result holds and that we get the state $|\varphi^{(m)}\rangle$ after $n < N$ swappings. It is simple to show from Eq. (1.19) that a ZZ measurement on $|\varphi^{(m)}\rangle \otimes |\varphi\rangle$ yields the state $|\varphi^{(m+1)}\rangle$ with probability $p_+^{(m)} \equiv (\varphi_0^{m+1} + \varphi_1^{m+1})/(\varphi_0^m + \varphi_1^m)$ and $|\varphi^{|m-1|}\rangle$ with probability $p_-^{(m)} \equiv 1 - p_+^{(m)}$. Then, let us calculate the variation of the SCP due to one swapping, considering that the set $\{(p_i, |\varphi^{(m_i)}\rangle), i = 1, \ldots, l\}$ describes the outcomes of the first n measurements. Denoting by $\lambda_\pm^{(m)}$ the smallest Schmidt coefficients of $|\varphi^{|m\pm 1|}\rangle$, the new SCP reads

$$S^{(n+1)} = \sum_{i=1}^{l} p_i \, 2\Big(p_+^{(m_i)} \lambda_+^{(m_i)} + p_-^{(m_i)} \lambda_-^{(m_i)}\Big)$$
$$= S^{(n)} - (\varphi_0 - \varphi_1)\, p_n^{(0)}, \quad (1.39)$$

where $p_n^{(0)}$ stands for the probability of getting the state $|\varphi^{(0)}\rangle = |\Phi^+\rangle$ after n measurements. Since this probability is non null for odd n only, it results that the SCP decreases after every second swapping only. In the case $n = 1$, this is the "conservation of entanglement" pointed out in Prop. 1.2. The quantity $p_n^{(0)}$ is calculated from Fig. 1.5: it is the weighted sum over all possible paths Γ that go from the root of the oriented tree to the node $m = 0$ at position n. We notice that the path weights, denoted by w_n, depend on n only and not on the paths themselves. This is indeed the fact since $p_+^{(m)} p_-^{(m+1)} = \varphi_0 \varphi_1$ for all m, and because the paths Γ start and terminate at the same level $m = 0$. Thus, for odd n we have $w_n = (\varphi_0 \varphi_1)^{(n+1)/2}$, and using basic combinatorial analysis one finds $p_n^{(0)} = \binom{2k}{k}(\varphi_0 \varphi_1)^k$, with $k = \frac{1}{2}(n+1) \in \mathbb{N}$. Finally, the average SCP after N entanglement swappings in the ZZ basis reads

$$S^{(N)} = 1 - (\varphi_0 - \varphi_1) \sum_{k=0}^{[N/2]} \binom{2k}{k} (\varphi_0\varphi_1)^k, \qquad (1.40)$$

where $[x]$ denotes the integer part of x. This last equation behaves as $(4\varphi_0\varphi_1)^{N/2}/\sqrt{N}$ for N tending to infinity, and consequently, measurements in the ZZ basis yield much better results than the XZ strategy.

CHAPTER 2

Long-distance entanglement in planar graphs

The previous chapter concluded with the impossibility of generating long-distance entanglement in a chain of non-maximally entangled pairs of qubits. In fact, the optimum measurements lead to an exponentially small probability of success with the number of required entanglement swappings, see Sec. 1.3. However, extending the quantum network to a system of higher (spatial) dimension grandly impacts on this probability. In this chapter, we indeed show that a Bell pair can be generated over a long distance in quantum networks with the topology of planar graphs.[1] More precisely, we propose some measurement strategies which yield a strictly positive probability of entangling two arbitrarily distant qubits, provided that the entanglement $S(\varphi)$ of the elementary connections $|\varphi\rangle$ is larger than a critical value S_c. These strategies are of two kinds, deterministic or purely statistical. While the former makes use of distillation methods, the latter exploits ideas of percolation theory. Both strategies are based on the same (and simple) observation, nevertheless: there exist plenty of weakly entangled paths between two distant nodes in a two-dimensional lattice, but only one chain of singlets is necessary to generate a Bell pair between these two nodes.

In Sec. 2.1, we apply the results of the previous chapter, in particular optimal entanglement swappings and distillation procedures, to two-dimensional networks of large size. We start by considering some *hierarchical* graphs, that is, networks that iterate certain geometric structures, so that at each level of iterations the number of nodes, or the number of neighbors, changes (Sec. 2.1.1). In particular, we discuss the "diamond" graph, for which we prove that for sufficiently large initial entanglement, one can establish perfect entanglement on large scales (*i.e.*, on some lower levels of iteration) in a finite number of steps. A similar result holds for a double binary tree, in which each iteration step branches every bond into two. For such graphs, if the initial entanglement is large enough, perfect entanglement can be established at each level of iteration. Then, in Sec. 2.1.2, we turn to genuine two-dimensional lattices. Using distillation techniques, we show that, for the square lattice, one can convert the connections of a sufficiently broad strip into a backbone of perfect singlets.

Percolation strategies for infinite lattices are considered in Sec. 2.2. For instance, we reconsider the hexagonal lattice with double bonds described in [ACL07] and then discuss the case of a triangular lattice with distinct bonds. In the first of these examples, quantum measurements lead to a local reduction of the SCP but change the geometry of the lattice, which increases

---

[1] Basics of graph theory are supposed to be known by the reader and can be found in any textbook. For instance, we refer the reader to [Die05] for a very good but easily accessible book on this topic.

its connectivity and thus the *classical entanglement percolation* threshold corresponding to a straight conversion into singlets of all the links of the lattice. We call this effect *quantum entanglement percolation*. In the second example, we use the measurements optimizing the SCP to transform the original lattice into a double-size triangular lattice with a higher probability of getting a singlet on the bonds. We also describe a different type of strategies, where a square lattice is transformed into two disjoint square lattices of double size but with the same average SCP. In this case, we prove that the probability of connecting a pair of neighboring nodes to another such pair is strictly larger than in the original protocol.

While the advantage of quantum entanglement percolation over the classical protocols has been found for some specific lattices only (Sec. 2.2), we propose in Sec. 2.3 a strategy that beats the classical entanglement percolation for all the lattices that we considered. This strategy is based on the creation of multipartite entangled states, see Sec. 2.3.1. It improves not only the entanglement threshold, but also the success probability of the protocol for any amount of entanglement in the connections (Sec. 2.3.3).

Finally, in Sec. 2.4, we briefly discuss the optimality of the various protocols proposed for generating a long-distance entangled pair of qubits. In particular, while any percolation threshold defines a *sufficient* condition, it is still an open and very interesting question to determine whether or not there exists a *necessary* amount of entanglement to achieve this task.

2.1 Deterministic methods

In this section, we consider graphs in which entanglement can be generated between arbitrarily distant nodes by making use of entanglement swappings and distillations. In fact, for some graphs and if the entanglement of the connections is large enough, it is possible to compensate the loss of entanglement due to the swappings with an entanglement concentration procedure. Clearly, the distillations require that several pairs of entangled qubits are created between two nodes, which is one of the basic reasons why graphs spanning the plane are so interesting in comparison to one-dimensional systems.

2.1.1 Hierarchical graphs

Let us first study the generation of entanglement over large scales in graphs that have a hierarchical geometry. These are graphs that iterate certain geometric structures, so that at each level of iteration the number of nodes (or neighbors) changes. Unfortunately, optimal strategies are not known for such graphs; we restrict our considerations to showing that one can generate perfect entanglement in a finite number of steps at some iteration level. This entanglement is further swapped to the lowest levels of iterations, *i.e.*, to the largest scales, which can be considered as the largest geometrical distances.

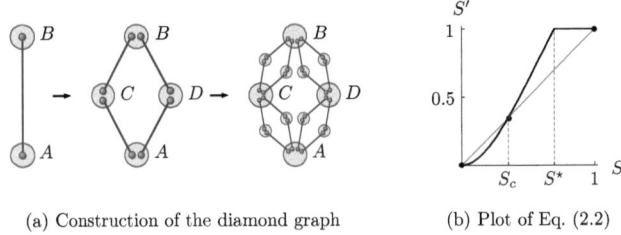

(a) Construction of the diamond graph (b) Plot of Eq. (2.2)

Figure 2.1: (a) The first two iterations of the diamond graph. (b) Recursion relating S on the higher level of iteration to S' at one lower level of iteration.

"Diamond" graph

In this section, we consider the so-called "diamond" graph, which is obtained by iterating the following operation, see also Fig. 2.1a: a single bond (one entangled state of two qubits) is replaced by four bonds forming a diamond shape. After k iterations, the nodes A, B, C, D have 2^k links, the nodes on the next level 2^{k-1}, etc. We now prove that for sufficiently large initial entanglement, one can establish Bell pairs on large scales in a finite number of steps.

We assume that the graph is formed by many iterations and that all bonds correspond to the entangled state $|\varphi\rangle$. Our aim is to perform some measurements in a recursive way, showing that, for sufficiently high $S(\varphi)$, it is possible to generate perfect entanglement on the lowest level of the iteration hierarchy, that is, between the "parent" nodes A and B. In order to keep the form of the network unchanged by the recursive measurements, we apply the WCE strategy to the nodes analogue to C and D, starting from the highest iteration level. The two entanglement swappings in the XZ basis yield, with unit probability, two identical pairs of entangled states which can then be distilled into a state $|\psi\rangle$ with larger entanglement. Remark that we are exactly in the situation of the single square network discussed in Sec. 1.2.2. From Eqs. (1.20) and (1.29), one finds

$$\psi_0 = \max\left\{\frac{1}{2}, \frac{1}{4}\left(1 + \sqrt{1 - (4\varphi_0\varphi_1)^2}\right)^2\right\}, \qquad (2.1)$$

and denoting the SCP of $|\varphi\rangle$ and $|\psi\rangle$ by S and S', respectively, we write the recursion relation as

$$S' = 1 + \frac{S^2(2-S)^2}{2} - \sqrt{1 - S^2(2-S)^2}, \qquad (2.2)$$

which has to be smaller than or equal to one, of course. This recursion has one unstable fixed point $S_c \approx 0.349$ and two stable fixed points $S = 0$ and $S = 1$, see Fig. 2.1b. The latter is achieved in a finite number of steps, provided that the initial entanglement satisfies $S > S_c$. Note that S_c is strictly smaller than $S^\star = 2(1 - \varphi_0^\star) \approx 0.7$, see Eq. (1.33), and that for $S \geq S^\star$ the maximum entanglement $S' = 1$ is achieved in one step only.

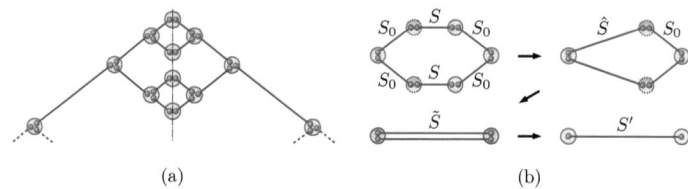

Figure 2.2: (a) Two binary trees facing each other. (b) Sequence of XZ measurements and distillation leading to a recursion relation for the SCP of the bonds.

Double binary tree

Similar results hold for the graph union of two binary trees, also called 3-Cayley trees, whose leaves are joined in the center, see Fig. 2.2a. Let us denote the initial SCP of all bonds $|\varphi\rangle$ by S_0 and perform a WCE measurement at all nodes in the middle of the tree. This prepares two two-qubit states between the neighboring nodes, which can be distilled into the pair of qubits $|\psi\rangle$ given in Eq. (2.1). We now describe the recursion strategy, see also Fig. 2.2b: first, an entanglement swapping in the XZ basis is performed at one of the middle nodes, which yields a state with SCP equal to $\hat{S} = S_{XZ}(S_0, S) \equiv 1 - \sqrt{1 - S_0(2 - S_0)S(2 - S)}$. Then, the WCE swapping is applied to the remaining pair of bonds, leading to $\tilde{S} = S_{XZ}(\hat{S}, S_0)$. Finally, the optimum entanglement distillation is applied to the pair of $\tilde{S}$ bonds obtained from the two different (but neighboring) branches of the tree. The recursion relation reads:

$$S' = S'(S, S_0) = \min\left\{1, 2\left(1 - (1 - \tilde{S}/2)^2\right)\right\}. \quad (2.3)$$

This recursion depends explicitely on S_0, and three distinct cases have to be distinguished, as depicted in Fig. 2.3:

(i) $S_0 < S_c$: if the derivative of the recursion function is smaller than unity at the origin, then there exists only the trivial (and stable) fixed point $S = 0$. Explicitly, S_c is found by solving the equation

$$\left.\frac{d}{dS}\right|_{S=0} S'(S, S_c) = 2\left(S_c(2 - S_c)\right)^2 = 1 \quad \Rightarrow \quad S_c \approx 0.459.$$

Therefore, for small values of S_0, entanglement cannot be generated over a large geometrical distance.

(ii) $S_c < S_0 < S^*$: in that case, one stable and non-trivial fixed point appears. Remark that this fixed point depends on S_0 and can reach the whole interval $]0, 1[$. The value S^* is found by solving $S'(1, S^*) = 1$ and is exactly the same as for the diamond graph: $S^* \approx 0.7$.

(iii) $S_0 > S^*$: if the entanglement of the bonds is large enough, then a Bell pair between distant nodes is generated in a finite number of iteration steps.

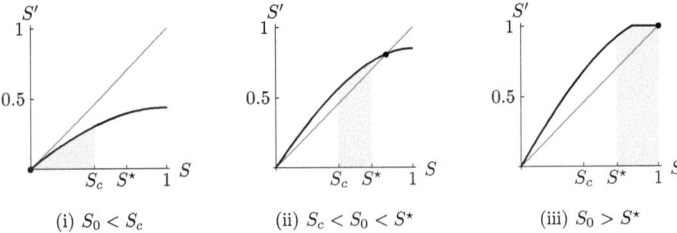

Figure 2.3: Recursion relation for the SCP of the double binary tree, determined by Eq. (2.3). Depending on S_0, three cases have to be distinguished. The fixed point of the recursion is highlighted by a bullet.

Note that these considerations can be qualitatively understood from the discussion of the diamond graph, since the two constructions are quite similar and because the WCE does not increase the SCP.

2.1.2 Regular lattices

In the previous section, we have shown that entanglement can be generated over a large distance in graphs with a hierarchical structure, if the entanglement of the bonds if larger than a critical value S_c. The self-similarity of these graphs allows one to design natural sequences of entanglement swappings and distillations but suffers a physical limitation: either the length of the bonds or the number of qubits per node grows exponentially with the iteration depth. From now on, we therefore consider regular two-dimensional *lattices*, that is, periodic configurations of nodes throughout the plane that have a finite number Z of neighbors. In what follows, we describe a deterministic strategy for lattices which have a coordination number Z strictly larger than three. In particular, we show how two infinitely distant nodes can be entangled with unit probability in the square lattice ($Z=4$), provided a sufficiently high entanglement in the connections. The generalization to other lattices of high connectivity, such as the triangular lattice ($Z=6$), is straightforward.

A "centipede" in the square lattice

As another example of the recursive measurement method developed in the previous section, we consider a wide stripe of a square lattice and the "centipede" figure within, see Fig. 2.4a. Let us denote the initial entanglement by S_0 and the entanglement at the end bond of a "leg" by S. We sequentially shorten the legs of the centipede, gradually concentrating the entanglement of the links such that we eventually get a perfect singlet at its "spine". Concretely, we perform on the extremity of each leg the XZ measurements depicted in Fig. 2.2b, with the difference that one of the paths has only one bond S_0 and not three. The last step of the iterative procedure is then to distill two states of entanglement S_0 and $\tilde{S} = S_{\mathrm{XZ}}(\hat{S}, S_0)$, with $\hat{S} = S_{\mathrm{XZ}}(S_0, S)$, yielding

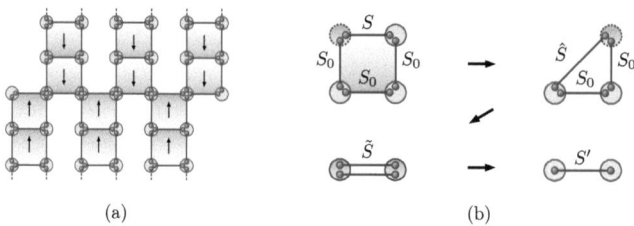

Figure 2.4: (a) "Centipede" with its "legs" and "spine". Remark that some bonds are not used. (b) Recursive measurement scheme; this method can be applied in higher dimensions too.

the recursion relation
$$S'(S, S_0) = \min\left\{1,\ S + \tilde{S} - \frac{S\tilde{S}}{2}\right\}. \qquad (2.4)$$

This situation is very similar to the case of the double binary tree, but there is a small difference, nevertheless. In fact, the current recurrence relation depends explicitly on S_0 and has always a non-trivial stable fixed point. This fixed point, however, is strictly smaller than unity when S_0 is small. In this case, although we do concentrate more entanglement along the spine of the centipede, we still have to face the problem that the spine is a one-dimensional system and thus exhibits an exponential decrease of the SCP with the number of entanglement swappings that are required. On the other hand, if S_0 is larger than the critical value $S_c \approx 0.649$, then the fixed point is a maximally-entangled state. This point is reached in a finite number of iteration steps; note that S_c is simply the smallest value such that $S'(1, S_c) = 1$.

The construction of the centipede, as proposed in Fig. 2.4a, implies that only the nodes lying on the (one-dimensional) spine can get entangled. It may therefore seem a huge waste of physical resources, since the ratio between the number of final Bell pairs and the consumed ones approaches zero for a lattice size tending to infinity. However, this observation is pertinent only close to the critical point S_c. In fact, for sufficiently large entanglement in the initial bonds, the width of the centipede, which is equal to twice the length of its legs, is very small. For instance, one can check that only one iteration step is necessary to get a maximally entangled pair if $S_0 \geq S^* \approx 0.684$, which is the solution of $S'(S^*, S^*) = 1$. In that case, many narrow centipedes can be created and, more important, connected together, as depicted in Fig. 2.5a. It follows that the resulting "giant" centipede spans a non-vanishing proportion of all the nodes, which tends to one third in the limit of infinite lattice size. Remark that this ratio is further increased to one-half by considering a slightly different measurement pattern, see Fig. 2.5b.

2.2 Strategies based on bond percolation

The strategies that have been proposed so far to achieve long-distance entanglement in two-dimensional networks are based solely on two properties of the pure states. First, entanglement

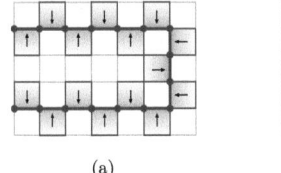 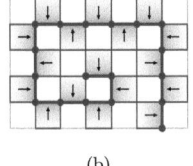

(a) (b)

Figure 2.5: (a) Horizontal centipedes of finite width can be connected at the borders of the lattice to form a giant centipede spanning one third of all the nodes (in the limit of infinite lattice size). (b) This proportion is raised to one-half by considering a spiral construction. In this case, all links of the lattice are used.

can be propagated (at high cost) through the lattice with the help of entanglement swappings. This is equivalent to performing a sequence of teleportations having each a non-unit probability of success: one station prepares locally an entangled pair of qubits, one of which is sent from node to node until it reaches its destination. The second ingredient is that multiple connections between two nodes can be distilled, compensating (or even surpassing) the loss of entanglement due to the swappings. Thus, the minimum requirement for generating long-distance entanglement with this strategy is that a "backbone" of Bell pairs is created in the lattice. This is possible if the initial entanglement of the bonds is larger than a critical value S_c and if the nodes that belong to the backbone have at least four neighbors each: two connections are part of the backbone, while the other two are used to distill the former. Since deterministic strategies typically create a one-dimensional chain of Bell pairs by using only a stripe of finite width from the lattice, they do not seem to exploit the full potentiality of two-dimensional networks. These considerations raise at least two important questions: how can one deal with lattices of small connectivity (as the honeycomb lattice for example), and can the critical value S_c of the deterministic methods be lowered by using other strategies? An affirmative answer to these questions was given [ACL07], bringing ideas of percolation theory to the context of quantum communication networks.[2]

In Sec. 2.2.1, we present the basics and some well-established results of classical percolation theory and recall the entanglement percolation protocols proposed in [ACL07]. Those are of two types: first, the so-called *classical entanglement percolation* (CEP) protocols, in which the quantum links $|\varphi\rangle$ of the lattices are converted into singlets with optimum probability $p = S(\varphi)$. This defines a (classical) percolation problem where bonds are inserted with probability p in the lattice. Second, in Sec. 2.2.2, we develop the ideas of [ACL07] showing that CEP strategies can be surpassed, in some cases, by *quantum entanglement percolation* (QEP) protocols. In this

---

[2]Note that the probabilistic nature of quantum physics makes percolation theory a particularly well-adapted toolbox for the study of quantum systems subject to measurements for instance. Ideas of percolation theory are for example useful in the context of quantum computing with non-deterministic quantum gates [KRE07]. The interested reader is referred to pp. 287–319 in [SBC09] for an overview of the application of percolation methods to the field of quantum information.

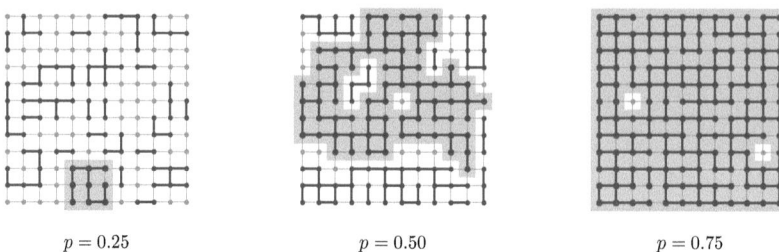

$p = 0.25$ $\quad\quad\quad\quad\quad\quad$ $p = 0.50$ $\quad\quad\quad\quad\quad\quad$ $p = 0.75$

Figure 2.6: Classical bond percolation: bonds are open (present) with probability p and closed otherwise, and groups of nodes connected by open bonds are called clusters. In these examples, the largest cluster is highlighted: typically, it is small and bounded for $p < p_c$ but spans a finite proportion of the lattice for $p > p_c$ (the critical value p_c is equal to 50% in infinite square lattices). In the latter case, the largest cluster is written $\mathcal{C}$ and is called the *giant cluster*.

scenario, joint measurements are previously performed at the nodes of the lattice. In particular, the fact that the average SCP does not decrease after one entanglement swapping allows one to define a bond percolation problem in a new lattice that has a favorable geometry. Such structural transformations have been further elaborated in [LWL09] and expanded to complex networks [CC09] and to bipartite mixed states [BDJ09].

2.2.1 Classical entanglement percolation

Classical percolation is perhaps one of the most fundamental examples of critical phenomenon, since it is a purely statistical one [Gri99]. At the same time, it is very universal, since it describes a whole variety of physical, biological or ecological processes; see [SA91] for a nice introduction to percolation theory. In bond percolation, we typically consider a regular lattice of nodes connected by random bonds, the probability of having a bond between two neighboring nodes being p, see Fig. 2.6. For a given infinite lattice, one would like to know whether an infinite open cluster exists, that is, whether there is a path of connected nodes of infinite length through the lattice. It turns out that an unique *giant cluster* appears if, and only if, the connection probability p is larger than a critical threshold p_c, which depends on the lattice. Few lattices have a threshold that is exactly known; among them we find the important honeycomb, square, and triangular lattices [SE64]:

$$p_c^{\bigcirc} = 1 - 2\sin(\pi/18), \quad p_c^{\square} = 1/2, \quad \text{and} \quad p_c^{\triangle} = 2\sin(\pi/18).$$

In this chapter, we extensively use such results of percolation theory, and we refer the reader to the above-cited books or to [BR06] for mathematical definitions and rigorous proves.

Suppose that we want to generate entanglement between two distant stations A and B in a quantum lattice, where each connection denotes a non-maximally entangled state $|\varphi\rangle$. Classical entanglement percolation runs as follows [ACL07]: every pair of neighboring nodes optimally converts its two-qubit state $|\varphi\rangle$ into a Bell pair, which succeeds with probability $S(\varphi)$, see Eq. (1.9). If this value is larger than the threshold p_c of the lattice, i.e., if $S > S_c \equiv p_c$, then a giant cluster $\mathcal{C}$ appears, and the probability that both A and B belong to $\mathcal{C}$ is strictly positive. In that case, one can find a path of singlets between the two nodes and perform the necessary entanglement swappings such that A and B become entangled. Note that this path is randomly generated by the measurement outcomes at the nodes, which contrasts with the deterministic methods described in the previous section.

At this point, let us be a little more precise and define the quantities characterizing long-distance entanglement. A quantity of primary interest, very related to the percolation threshold, is the *percolation probability*

$$\theta(p) \equiv P(A \in \mathcal{C}), \qquad (2.5)$$

being the probability that a given node A belongs to the giant cluster. Clearly, $\theta(p) = 0$ for $p < p_c$, while $\theta(p) > 0$ for $p > p_c$. In quantum communication, we are interested in the probability $P(A \multimap B)$ of creating a Bell pair between two nodes A and B separated by a distance L. For $p < p_c$, this probability decays exponentially with $L/\xi(p)$, where the *correlation length* $\xi(p)$ describes the typical radius of an open cluster. Above the critical point, the two nodes are connected only if they are both in $\mathcal{C}$. In the limit of large L, the events $\{A \in \mathcal{C}\}$ and $\{B \in \mathcal{C}\}$ are independent, which, together with translational invariance, reduces the problem to studying $\theta(p)$:

$$P(A \multimap B) = \theta^2(p). \qquad (2.6)$$

2.2.2 Quantum entanglement percolation

A natural question one may ask is whether the thresholds S_c defined by classical percolation theory are optimal. In fact, entanglement percolation represents a related but different theoretical problem, where new bounds may have to be obtained. This is of course equivalent to determine if the measurement strategy based on local SCP is optimal in the asymptotic regime. Here, we construct several examples that go beyond the classical percolation picture, proving that the classical entanglement percolation strategy is not optimal. The key ingredient for the construction of these examples is the measurement strategy obtained for the one-swapper configuration that maximizes the SCP, namely the entanglement swapping in the ZZ basis (Prop. 1.2).

Honeycomb lattice with double bonds

The first example, which was already discussed in [ACL07], considers a honeycomb lattice where each pair of neighboring nodes is connected by two copies of the same two-qubit state $|\varphi\rangle$, see Fig. 2.7a. The CEP strategy converts in the best possible way all bonds shared by two parties into one singlet. Majorization theory tells us how much entanglement can be distilled from each double bond in case of weak entanglement, namely $S(|\varphi\rangle^{\otimes 2}) = 2(1 - \varphi_0^2)$, see Eq. (1.29). We choose this conversion probability to be equal to the percolation threshold for the honeycomb

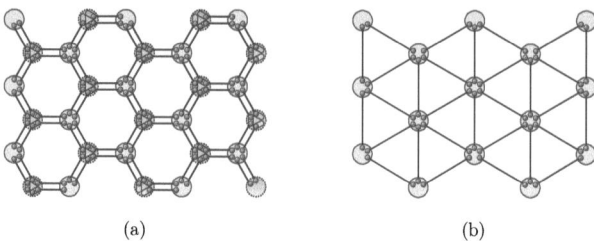

Figure 2.7: Each bond of the honeycomb lattice consists of two copies of the state $|\varphi\rangle$. (a) The dashed nodes perform three optimal entanglement swappings according to the SCP. (b) A triangular lattice of identical average SCP is obtained, and classical entanglement percolation is now possible in this new lattice.

lattice, which results in an entanglement threshold S_c for the state $|\varphi\rangle$:

$$S_c = 2\left(1 - \sqrt{1 - \frac{p_c^\bigcirc}{2}}\right) = 2\left(1 - \sqrt{\frac{1}{2} + \sin\left(\frac{\pi}{18}\right)}\right) \approx 0.358. \quad (2.7)$$

Now, we show that there exists a QEP strategy yielding a better percolation threshold: half the nodes perform on their qubits three entanglement swappings in the ZZ basis, mapping the honeycomb lattice into a triangular lattice as depicted in Fig. 2.7b. The important ingredient of this construction is that the SCP of the new bonds is exactly the same as the SCP of the initial state $|\varphi\rangle$. We set its entanglement to be $p_c^\triangle$, and a lower threshold is found:

$$S_c = p_c^\triangle = 2\sin\left(\frac{\pi}{18}\right) \approx 0.347. \quad (2.8)$$

This proves that CEP is not optimal since it cannot generate long-distance entanglement if $0.347 < S(\varphi) < 0.358$, while the proposed QEP strategy achieves it with a strictly positive probability.

Asymmetric triangular lattice

A second example, although less symmetric, is generic and has a totally different character than the previous one. For simplicity, we show the argument in the case of a triangular lattice, but the same reasoning can be applied to other geometries. Consider the asymmetric triangular lattice depicted in Fig. 2.8a. A solid line corresponds to the two-qubit pure state $|\varphi\rangle$, while a dashed line represents a less entangled state $|\tilde{\varphi}\rangle$, i.e., $S > \tilde{S}$ (with obvious notation). We choose the first state such that its amount of entanglement satisfies $p_c^\triangle < S < \sqrt{p_c^\triangle}$. If $|\tilde{\varphi}\rangle = |\varphi\rangle$, the classical entanglement percolation strategy works. However, we tune the entanglement of the second state such that its SCP is small enough to make the classical entanglement percolation impossible. Such a tuning always exists. In fact, we note that when $\tilde{S} \to 0$, these states can simply be removed from the lattice, and classical entanglement percolation fails since we have

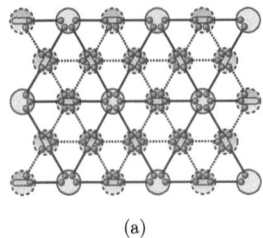 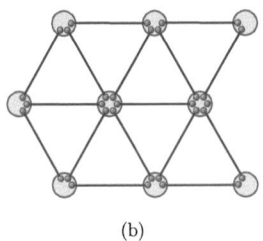

Figure 2.8: An asymmetric triangular lattice. (a) The lattice consists of two different entangled states $|\varphi\rangle$ and $|\tilde{\varphi}\rangle$, drawn in solid and dashed lines, respectively. The less entangled states $|\tilde{\varphi}\rangle$ are discarded, and some of the nodes perform an entanglement swapping in the ZZ basis. (b) A new triangular lattice is obtained, solely governed by the SCP of $|\varphi\rangle$.

$S^2 < p_c^\triangle$ and because the links of the new lattice consists of two consecutive states $|\varphi\rangle$. It is now straightforward to construct a successful QEP strategy: the states $|\tilde{\varphi}\rangle$ are discarded and the optimal SCP strategy for one entanglement swapping is performed, see Fig. 2.8b. The lattice is transformed into another triangular lattice, in which all links have the same amount of entanglement $S > p_c^\triangle$, so that CEP can now be successfully applied.

Doubling the square lattice

The final example has yet another character and deals with a square lattice in which all bonds are identical and described by the state $|\varphi\rangle$. Here, we replace every second pair of horizontal bond by a single one performing an entanglement swapping in the ZZ basis; let us recall that this operation does not decrease the average SCP of the bonds (Prop. 1.2). The same is done with every second pair of vertical bonds, and we consequently replace the original square lattice by two disjoint lattices having a lattice constant that is twice larger than the original one, see Fig. 2.9. Now, we are interested in establishing long-distance entanglement between A or A' and B or B'. The nodes A and A' (B and B') are next-nearest neighbors in the original lattice but belong to different double-size lattices.

Let us first consider the QEP strategy for which calculations are simple. The pairs (A, B) and (A', B') belong to two disjoint lattices, and the probability that one of these pairs belongs to the percolating cluster is asymptotically equal to $\theta^2(p)$, with $p = S(\varphi)$. The probability that at least one of the pairs belongs to the giant cluster is thus

$$P_{\text{QEP}} = 1 - (1 - \theta^2)^2 = \theta^2 (2 - \theta^2). \tag{2.9}$$

This probability has to be compared with the CEP probability that at least one of the pairs (A, B), (A', B), (A, B') or (A', B') belongs to the giant cluster in the original square lattice. Asymptotically, it is given by $P_{\text{CEP}} = \pi^2$, where π is the probability that A or A' (or

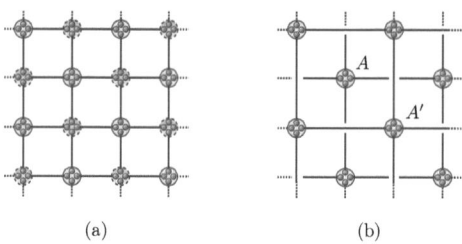

(a) (b)

Figure 2.9: (a) Measurements doubling the lattice constant of the square lattice: the dashed nodes perform an entanglement swapping along the vertical and horizontal directions. (b) Two disjoint square lattices with double lattice constant are created, and we want to establish an entangled pair of qubits between A or A' and B or B' (not shown).

equivalently B or B') belongs to the giant cluster:

$$\pi = P(A \vee A' \in \mathcal{C}) = P(A \in \mathcal{C}) + P(A' \in \mathcal{C}) - P(A \wedge A' \in \mathcal{C}). \tag{2.10}$$

In order to estimate the last term in the above expression, we use the celebrated Fortuin–Kasteleyn–Ginibre (FKG) inequality [FKG71]. To that end, we first recall the following definition: an event described in terms of a percolation configuration is said to be *increasing* if it has the property that, once it holds for a certain bond configuration, it holds for all configurations obtained by adding bonds to the initial one. The FKG inequality then states that any two such events are positively correlated. The events $\{A \in \mathcal{C}\}$ and $\{A \circ\!\!-\!\!\circ A'\}$ ("A and A' are connected by a path of maximally entangled bonds") are clearly increasing, and since their intersection is the event $\{A \wedge A' \in \mathcal{C}\}$, it follows that

$$P(A \wedge A' \in \mathcal{C}) \geq P(A \in \mathcal{C}) \, P(A \circ\!\!-\!\!\circ A'). \tag{2.11}$$

Denoting $P(A \wedge A' \in \mathcal{C})$ by τ, we verify that $\pi \leq \theta(2 - \tau)$, and therefore doubling the square lattice is a better strategy than the classical percolation, *i.e.*, $P_{\mathrm{QEP}} \geq P_{\mathrm{CEP}}$, whenever

$$(2 - \tau)^2 \leq 2 - \theta^2. \tag{2.12}$$

Let us show that this inequality holds when p lies just above the threshold $p_c^{\square} = 0.5$. In that case, θ tends to zero, and we thus have to show that $\tau \geq 2 - \sqrt{2} \approx 0.586$. We may try to estimate τ from below by considering the six shortest trajectories connecting A and A': the most direct ones (two paths of length two), and the two pairs of four-bond paths around the adjacent squares. This leads to $\tau > 2q - q^2$, with $q = p^2 + 2p^4 - 2p^5$. Unfortunately, for $p = 0.5$, this estimate is too small and thus not sufficient to our purpose. In fact, it gives only $\tau > 0.527$. One can improve the estimate analytically by adding further paths connecting A and A'. This procedure becomes, however, technically tedious, and we therefore turn to a standard numerical Monte Carlo simulation for calculating the transition probability from A to A'. The method we use generates the shortest paths (like the ones used for calculating the above estimate) automatically, while the longer ones are introduced by the Monte Carlo sampling. For $p > p_c^\square$, the convergence is exponential: if we plot a subsequent estimate of τ as a function of the maximum cluster size allowed in the Monte Carlo sampling, it approaches the final value exponentially fast for large clusters. As expected, the convergence is algebraic at $p = p_c^\square$: the estimate of τ approaches its final value as a power of the cluster size. A power law fit and a comparison with the values just above the percolation threshold give, with a very good accuracy, $\tau \simeq 0.687$. This proves that quantum mechanical measurements can improve the classical percolation strategy. In this example, the QEP protocol does not lead to a better threshold but yields a higher probability of generating a long-distance Bell pair for p approaching $p_c^\square$ from above. Note that the inequality $P_{\text{QEP}} > P_{\text{CEP}}$ also holds for all $p \in]p_c^\square, 1[$, as recently shown in Sec. IIIC of [LWL09].

2.3 Multipartite entanglement percolation

In the previous section, we have shown how CEP can be enhanced by previously applying some quantum operations at the nodes. All examples given so far consist of transforming the quantum lattices by a sequence of entanglement swappings in the ZZ basis, thus conserving the average entanglement of the bonds. These examples are, however, restricted to pure geometrical transformations, and apply to some specific lattices only. In this respect, it is not clear whether or not the CEP strategy, and particularly the corresponding threshold, can be improved in the simple square or triangular lattices for instance.

In this section, we introduce a powerful class of QEP strategies that exploit multipartite entanglement. In contrast to the protocols designed in Sec. 2.2.2, which solely employ entanglement swappings and conversions into singlets, we make full use of both the classical and quantum aspects of a quantum network: connectivity of the nodes and multipartite entanglement, respectively. The interplay of geometrical lattice transformations and entanglement manipulations is in fact a key ingredient in surpassing CEP.

In Sec. 2.3.1, we generalize the usual entanglement swapping to the case of three or more qubits. This quantum operations naturally introduces the Greenberger-Horne-Zeilinger (GHZ) states

$$|\text{GHZ}_n\rangle \equiv \frac{|0\rangle^{\otimes n} + |1\rangle^{\otimes n}}{\sqrt{2}}, \qquad (2.13)$$

which are the generalization of the Bell pair $|\Phi^+\rangle$ to n qubits. Then, in Sec. 2.3.2, we describe how GHZ states can percolate in a quantum lattice, leading to a site rather than a bond percolation process. Finally, we show in Sec. 2.3.3 that this multipartite strategy systematically outperforms the CEP method, regardless of the initial entanglement of the bonds and of the lattice geometry. If in the previous section it was difficult to contrive an example where CEP

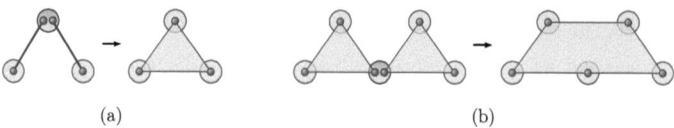

Figure 2.10: Generalized entanglement swapping. (a) Two partially entangled pairs of qubits are probabilistically converted into a GHZ state of three qubits. (b) Two (or more) GHZ states can be merged into a larger structure. Here, we depict the operation $|GHZ_3\rangle^{\otimes 2} \to |GHZ_5\rangle$.

can be surpassed, it is now hard to find an example that does *not* admit any improvement. Indeed, multipartite entanglement percolation leads to higher connection probabilities for all the lattices that were considered.

2.3.1 Generalized entanglement swapping

The basic quantum operation that creates entanglement between the extremities of two entangled pairs qubits is a measurement of the middle qubits, as discussed in Sec. 1.1.1 (see also Figs. 1.1a and 2.10a). The main and simple (but powerful) idea of this section is to replace the complete measurement at B by a generalized measurement $\mathcal{B}$ with operators

$$\begin{aligned}\mathcal{B}_0 &= |0\rangle\langle 00| + |1\rangle\langle 11|, \\ \mathcal{B}_1 &= |0\rangle\langle 01| + |1\rangle\langle 10|,\end{aligned} \quad (2.14)$$

which satisfy the completeness relation $\sum_{i=0}^{1} \mathcal{B}_i^\dagger \mathcal{B}_i = \mathbb{1}_4$. The second outcome leads to a perfect GHZ state among the three nodes (after a bit-flip X at C), but with probability $\varphi_0^2 + \varphi_1^2$ the first outcome yields the (unnormalized) state $\varphi_0|000\rangle + \varphi_1|111\rangle$. As for a pair of qubits, the latter state is transformed into $|GHZ_3\rangle$ with probability $2\varphi_1^2/(\varphi_0^2+\varphi_1^2)$. Summing the two possibilities, we find that two copies of $|\varphi\rangle$ are converted into one GHZ state of three qubits with optimal probability $2\varphi_1$. The multipartite method derives its power from the fact that, while the value of this probability is the same as for the two-qubit swapping used in previous entanglement percolation protocols, we now have three entangled qubits rather than two.

The above procedure can be generalized to construct GHZ states of $n+1$ qubits starting from n copies of $|\varphi\rangle$ sharing a common node. In this case, we apply 2^{n-1} measurement operators E_m of the form $|0\rangle\langle m| + |1\rangle\langle \overline{m}|$, where $\overline{m}$ is the complement of m written in base 2. With a little thought, one sees that this generalized measurement is equivalent to a sequence of $n-1$ entanglement swappings in the ZZ basis, exactly as for a chain of n partially-entangled states. It follows that the probability of creating a GHZ state of $n+1$ qubits is given by Eq. (1.40) with $N = n-1$, that is, $P(GHZ_{n+1}) = S^{(n-1)}$. Remark that this construction is optimal for $n=3$ only, and that higher success probabilities can be found by considering different measurement bases, as discussed in Sec. 1.2.1. Suppose now that two perfect GHZ states of size n and m have been created in the lattice, and that they share one common node, say B. One can then build a larger GHZ state on $n+m-1$ nodes with unit probability by performing the generalized measurement $\mathcal{B}$ on the two qubits of B and by applying some bit-flips X depending on the outcome (see Fig. 2.10b):

$$|\text{GHZ}_n\rangle \otimes |\text{GHZ}_m\rangle \xrightarrow[\text{prob}=1]{\text{LOCC}} |\text{GHZ}_{n+m-1}\rangle.$$

Finally, let us remark that, given a GHZ state of any size, a perfect Bell pair is created between any two of its qubits by measuring all other qubits in the X basis.

GHZ states, CEP, and classical pathfinding algorithms

Before turning to multipartite entanglement percolation, let us show that the creation of GHZ states can also be useful in the last step of the CEP protocols. In fact, after the conversion into singlets of the bonds of the lattice, we have to find a path between the two distant nodes A and B that want to share a Bell pair (we consider that both belong to the giant cluster). Then, every node lying on this path has to behave as an entanglement swapper. There exist efficient classical algorithms to find the shortest path between two nodes in a graph, as the one proposed by Dijkstra in [Dij59], but the time required to solve this problem scales with the size of the lattice. It follows that, right after the singlet conversion, the nodes do not know if they further have to perform an entanglement swapping on some of their qubits. Instead of waiting for this information to come, every node but A and B can apply a GHZ measurement on the qubits that belong to a successfully converted Bell pair. These local operations create a "giant" GHZ state spanning the lattice, which is done in a time that does not scale with the lattice size. Hence, quantum correlations between A and B are already available before the (time-consuming) classical processes occur, which may be favorable in the context of quantum key distribution for instance [GRTZ02]. Remark that the outcome results of all sites that belong to the giant cluster are still required (at some later time) to deduce which of the Bell states was used for creating the quantum correlations between A and B.

2.3.2 An illustrative example

Let us consider a honeycomb lattice whose bonds are given by the partially entangled state $|\varphi\rangle$. In what follows, we first show how the multipartite strategy improves the entanglement threshold, which is related to a natural *site percolation* process.[3] Then, we carefully describe how the efficiency of CEP is compared with that of the multipartite protocol.

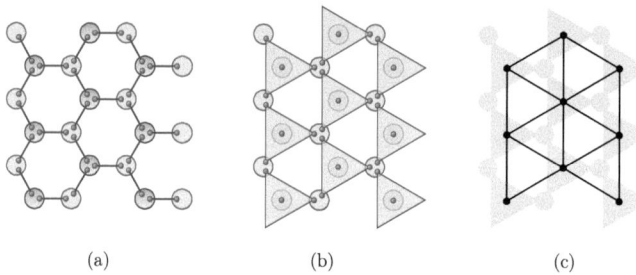

Figure 2.11: From bond percolation in the honeycomb lattice to site percolation in the triangular lattice. (a) In the honeycomb lattice $\mathcal{L}$, half the nodes apply a GHZ projection on their qubits. (b) GHZ states of four qubits (triangles) are created with probability $\hat{p}$ each. (c) The entanglement threshold is related to the site percolation threshold in a new lattice $\hat{\mathcal{L}}$ (here, the triangular lattice).

Entanglement threshold: from bond to site percolation

Each second node of the honeycomb lattice $\mathcal{L}$ performs a generalized entanglement swapping on its three qubits, as depicted in Fig. 2.11a, which creates a state $|\text{GHZ}_4\rangle$ with probability[4] $\hat{p} = P(\text{GHZ}_4)$. We then define a new lattice $\hat{\mathcal{L}}$, in which the vertices represent the GHZ states, and where two vertices are connected by a bond if the corresponding GHZ states share a common node in $\mathcal{L}$, see Figs. 2.11b and 2.11c. Adopting the usual terminology, we say that a site is open in $\hat{\mathcal{L}}$ if the GHZ transformation succeeded and closed otherwise. Since two GHZ states sharing one common node can be merged into a larger connected structure, it is easy to see that long-distance entanglement is possible if the corresponding site percolation leads to a giant cluster in $\hat{\mathcal{L}}$. In the present case, $\hat{\mathcal{L}}$ is the triangular lattice, whose site percolation threshold is equal to one-half, see [BR06] for instance. Therefore, a giant GHZ state appears if

$$\hat{p}(\varphi) > p_c^{\text{site}}(\hat{\mathcal{L}}). \qquad (2.15)$$

In the present example, performing the generalized entanglement swapping in the computational basis is not sufficient to beat the CEP threshold. In fact, since it corresponds to two consecutive swappings in the ZZ basis, one finds from Eq. (1.40) that the entanglement threshold $\hat{S}_c$ for this multipartite strategy, which is found by solving

$$\hat{p}(\varphi^*) = 6(\varphi_1^*)^2 - 4(\varphi_1^*)^3 = \frac{1}{2} = p_c^{\text{site}}(\triangle), \qquad (2.16)$$

is exactly equal to the CEP threshold S_c in the original lattice: $\hat{S}_c = 2\varphi_1^* = 1 - 2\sin(\pi/18) = p_c^{\text{bond}}(\bigcirc) = S_c$. It is in fact a pure coincidence that both $\hat{S}_c$ and $p_c^{\text{bond}}(\bigcirc)$ are solution of the

---

[3]Note that bond percolation can always be "artificially" mapped to site percolation.

[4]For clarity, all percolation parameters marked with a circumflex, such as $\hat{p}$ or $\hat{\theta}$, refer to multipartite entanglement protocols.

algebraic equation $1 - 3x^2 + x^3 = 0$, which, in the latter case, arises from the so-called star-delta transformation [SE63, SE64]. However, we know from Sec. 1.2.1 that the bases for the generalized entanglement swapping can be optimized, increasing the probability $\hat{p}$ of creating the GHZ states. For instance, optimizing these measurements over the Bell bases indeed yields $\hat{S}_c \approx 0.6182$, which is smaller than $S_c \approx 0.6527$. Finally, the entanglement threshold can be made even smaller by considering non-Bell measurements, see Tab. 1.1: a numerical optimization leads to $\hat{S}_c \approx 0.6090$, which clearly beats the CEP strategy.

Long-distance entanglement above the percolation threshold

The minimum amount of entanglement necessary to create a long-distance entangled pair is obviously a very interesting question, but it is also important to quantify the efficiency of the protocol for a given resource $|\varphi\rangle$; in what follows, we shall assume that its entanglement lies above the percolation threshold. As discussed in the previous sections, the quantity of interest is $P(A \circ\!\!\!-\!\!\!\circ B)$, which, for CEP, reduces to the study of the percolation probability $\theta(p)$ through the equation $P(A \circ\!\!\!-\!\!\!\circ B) = \theta(p)^2$. Since multipartite measurements change the topology of the underlying lattice, one has to define carefully the corresponding quantity $\hat{\theta}(\hat{p})$, such that $\hat{P}(A \circ\!\!\!-\!\!\!\circ B) = \hat{\theta}^2(\hat{p})$ holds for very distant nodes A and B. In fact, one cannot consider the usual percolation probability in the transformed lattice, because the new vertices are not directly related to the nodes of the original lattice. For this reason, it is clearer to consider the creation of the GHZ states in the original lattice, as depicted in Fig. 2.11b. We then define $\hat{\theta}(\hat{p})$ to be the probability that a node is connected to the giant GHZ state, or equivalently, that at least one of the GHZ connected to this node is successfully created and belongs to the giant cluster. In terms of the probability measure on the site percolation process on $\hat{\mathcal{L}}$, we have for a node A:

$$\hat{\theta}(\hat{p}) \equiv \hat{P}(\cup_i \{\hat{A}_i \in \hat{\mathcal{C}}\}), \tag{2.17}$$

where the union is over all the sites $\hat{A}_i$ possessing a qubit that is also in A. Note that $\hat{\theta}(\hat{p})$ now depends on the choice of A, since the nodes which did not perform any multipartite measurement are surrounded by three GHZ states, while the others are connected to one GHZ state only. For the multipartite strategy, we therefore restrict the two nodes to be chosen only from those at which no generalized entanglement swapping is to be made. More generally, both $\theta(p)$ and $\hat{\theta}(\hat{p})$ depend on the sites of the lattice, so that we consider only the nodes for which these values are maximum. We denote by f and $\hat{f}$ the fraction of such nodes in $\mathcal{L}$ and $\hat{\mathcal{L}}$, respectively, which allows a fair comparison between CEP and the multipartite protocols. In fact, the latter strategy always consumes some nodes for the creation of the GHZ states, and consequently we have $f < \hat{f}$, in general. Finally, the percolation probabilities are calculated by Monte Carlo simulations, showing that the multipartite strategy not only leads to a better threshold than CEP but is also favorable in the whole range $S \in [S_c, 1)$, see Sec. 2.3.3 (a different measurement pattern is used there, however).

Percolation probability close to unity

Let us consider the regime in which the connections are highly entangled, *i.e.*, $p = 2\varphi_1 = 1 - \varepsilon$ with $\varepsilon \ll 1$. In that case, clearly, we also have $\hat{p} = 1 - \hat{\varepsilon}$ with $\hat{\varepsilon} \ll 1$, but one can verify that $\hat{\varepsilon} > \varepsilon$ for any GHZ measurement involving more than two links. An analytical study of the percolation probabilities becomes possible in this regime of entanglement, and we use the perimeter method to compute high-density series expansions of $\theta(p)$ and $\hat{\theta}(\hat{p})$. The reader is referred for instance to App. C in [LWL09] for a description of this method, but for completeness let us explain it in the present situation. From Fig. 2.11b, it is clear that at the lowest order in $\hat{\varepsilon}$ one finds $\hat{\theta}(\hat{p}) = 1 - \hat{\varepsilon}^3 + \mathcal{O}(\hat{\varepsilon}^4)$. In fact, since a node that did not perform any measurement is connected to three GHZ states (those are created with probability $\hat{p}$ each), its probability not to be connected to the giant cluster is, at this order, given by $(1 - \hat{p})^3 = \hat{\varepsilon}^3$. One can further compute the next non-trivial order of the percolation probability by the following reasoning. Suppose that exactly one out of the three GHZ states is successfully generated. Then the node is disconnected from the giant cluster if the multipartite conversion failed for all neighboring nodes, which occurs with probability $(1 - \hat{\varepsilon})^6 \hat{\varepsilon}^6 \approx \hat{\varepsilon}^6$. Because this situation happens independently three times, we finally get

$$\hat{\theta}(\hat{p}) = 1 - \hat{\varepsilon}^3 - 3\hat{\varepsilon}^6 + \mathcal{O}(\hat{\varepsilon}^7). \tag{2.18}$$

2.3.3 The superiority of multipartite strategies

Now that the basic ingredients have been presented, we show that multipartite strategies outperform all previously known classical or quantum strategies, regardless of the initial entanglement of the bonds and for every lattice that we considered. From the above discussion on the percolation probability close to unity, a necessary condition for this statement to hold is that the measurement pattern does not comprise generalized entanglement swappings on more than two links. We hence focus on the measurements described in Eq. (2.14), such that $\hat{p} = p = 2\varphi_1$.

This section is structured as follows. First, we list the various lattices that are considered, and we carefully describe the measurement patterns that are applied on them. Second, we examine the thresholds and show that $\hat{p}_c < p_c$ for each lattice. Then, we compute the expansions $\hat{\theta}(p)$ and $\theta(p)$ in the high-density (*i.e.*, maximally-entangled) limit to prove that $\hat{\theta}(p) > \theta(p)$ as p tends to unity. Finally, using Monte Carlo techniques, we show that $\hat{P}(A \leftrightsquigarrow B) > P(A \leftrightsquigarrow B)$ for all $p > \hat{p}_c$.

Lattice transformations and percolation thresholds

We start by considering the $(4, 8^2)$ Archimedean lattice,[5] see [GS86] for this notation. In the first panel of Fig. 2.12, we propose a measurement pattern indicating on which qubits the generalized

---

[5] An *Archimedean lattice* is a tiling of the plane by regular polygons, in which each vertex is surrounded by the same sequence of polygons (here, one square and two octagons).

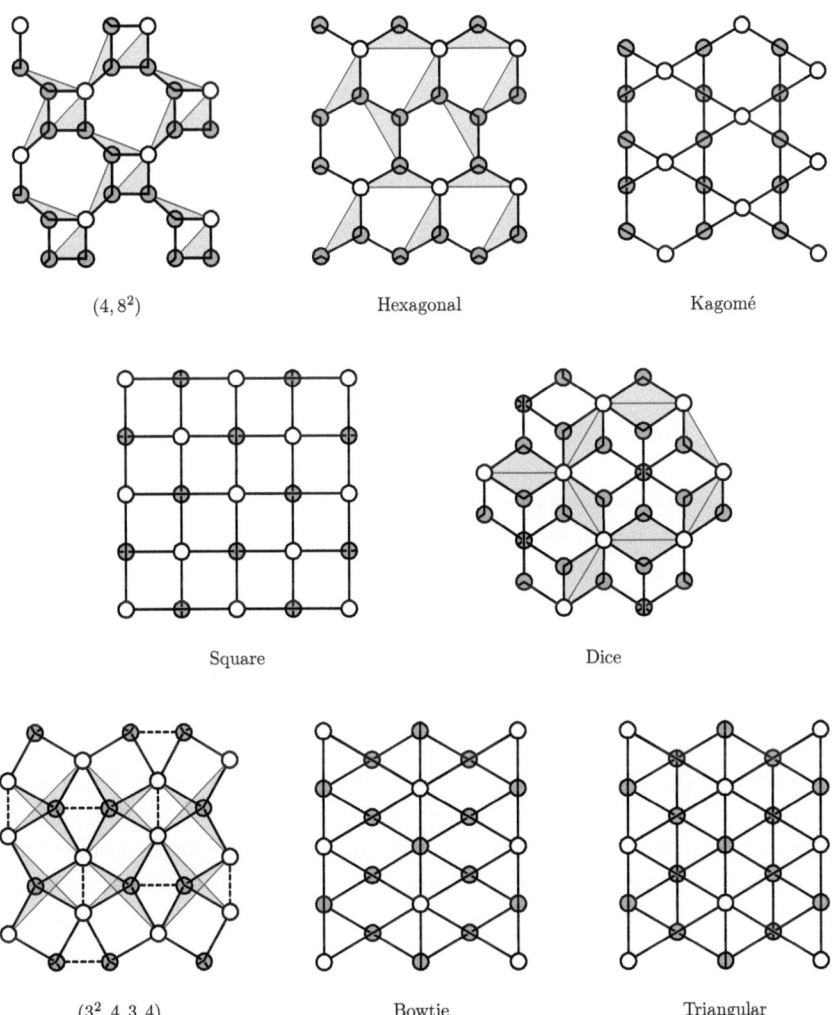

Figure 2.12: Lattice transformations under generalized entanglement swappings. Some nodes (filled circles) perform one or more GHZ measurements on their qubits, which are not drawn for clarity. Nodes where no quantum operation is performed (empty circles) are used for computing the percolation probability $\hat{\theta}$. Finally, in the $(3^2, 4, 3, 4)$ lattice, some of the links (dashed lines) are simply converted into Bell pairs as in CEP.

#	Lattice	p_c	$\hat{p}_c$	Δ [%]
1	$(4,8^2)$	0.6768	0.6499	4.0
2	Hexagonal	0.6527	0.609(0)	6.7
3	Kagomé	0.5244	0.427(1)	18.6
4	Square	0.5000	0.392(8)	21.4
5	Dice	0.4755	0.375(5)	21.0
6	$(3^2,4,3,4)$	0.4141	0.344(7)	16.8
7	Bowtie	0.4045	0.294(9)	27.1
8	Triangular	0.3472	0.273(5)	21.2

Table 2.1: Improvement of the entanglement percolation thresholds using a multipartite strategy, and relative gain $\Delta \equiv 1 - \hat{p}_c/p_c$. We performed Monte Carlo simulations to calculate $\hat{p}_c$ for the lattices 2–8. All other values are to be found, with higher precision, in [Gri99, Par04, NMW08].

entanglement swapping is applied: the new lattice is denoted by $^2\!/_3(3^2,6^2) + ^1\!/_3(3,6,3,6)$, and its critical point is $\hat{p}_c \approx 0.6499$ [NMW08]. Since the original threshold is $p_c \approx 0.6768$ [SZ99], the proposed strategy yields an improvement over CEP. It is interesting to note that the transformed lattice, arising from simple quantum operations, is among the rather exotic examples considered in other studies. For example, this two-uniform lattice is considered in [NMW08], where a quantitative relation between percolation thresholds and the Euler characteristic is demonstrated.

We studied many other lattices and found that multipartite entanglement percolation leads to better thresholds than CEP in every case, which suggests that it may be a universal result. All the lattices that were considered are depicted, together with the corresponding measurement patterns, in Fig. 2.12. Many such patterns lead to an improvement over the CEP strategy, but for simplicity we considered only periodic ones. For the square lattice, for instance, one can pair the links in such a way that two overlapping square lattices of double size are created. The key advantage of the multipartite strategy, with respect to the construction described in Fig. 2.9, is that we do not get disjoint lattices anymore, but rather connected ones since the middle qubits can be used to propagate the entanglement through the network.

We did not find published values of the site percolation thresholds in the resulting lattices, mainly because they are non-planar or non-regular graphs, and thus we turned to Monte Carlo simulations. The values obtained are summarized in Tab. 2.1: thresholds are not only better for all lattices, but the gain is often significant, especially for lattices of high connectivity.

Percolation probabilities above the thresholds

We have provided numerical evidences that multipartite entanglement percolation yields better thresholds than CEP, and therefore that the connection probability between two widely separated nodes is increased when the bond entanglement $p(\varphi)$ lies in the interval $(\hat{p}_c, p_c)$. In the

#	θ	$\hat{\theta}$	f
1	$1-\varepsilon^3-4\varepsilon^4-11\varepsilon^5$	$1-\varepsilon^3-4\varepsilon^4-4\varepsilon^5$	$1/4$
2	$1-\varepsilon^3-3\varepsilon^4$	$1-\varepsilon^3-\varepsilon^4$	$1/4$
3	$1-\varepsilon^4-6\varepsilon^6$	$1-\varepsilon^4-2\varepsilon^7$	$1/3$
4	$1-\varepsilon^4-4\varepsilon^6$	$1-\varepsilon^4-4\varepsilon^7$	$1/2$
5	$1-\varepsilon^6-6\varepsilon^7$	$1-\varepsilon^6-9\varepsilon^{10}$	$3/4$
6	$1-\varepsilon^5-5\varepsilon^8$	$1-\varepsilon^5-\varepsilon^8$	$1/2$
7	$1-\varepsilon^6-4\varepsilon^8$	$1-\varepsilon^6-4\varepsilon^{11}$	$1/2$
8	$1-\varepsilon^6-6\varepsilon^{10}$	$1-\varepsilon^6-2\varepsilon^{12}$	$1/4$

Table 2.2: Series expansions of $\theta(p)$ and $\hat{\theta}(p)$, for $p = 1 - \varepsilon$ and $0 \leq \varepsilon \ll 1$. These formulas are derived for a fraction $f = \hat{d}/d$ of nodes, where d ($\hat{d}$) is the density of nodes of higher connectivity in the original (transformed) lattice. Note that $d = 1$ for the Archimedean lattices, since by definition all their vertices are equivalent, while $d = 1/3$ for the dice and $d = 1/2$ for the bowtie lattices.

opposite regime in which connections are highly entangled, with $p = 1 - \varepsilon$ and $0 < \varepsilon \ll 1$, we use the perimeter method exactly as in the example of Sec. 2.3.2 to prove that the percolation probability $\hat{\theta}(p)$ is strictly larger than $\theta(p)$. The two lowest non-trivial orders of the expansions are easily calculated and are sufficient to show that the multipartite strategy leads to an improvement over CEP for all lattices, see Tab. 2.2. Interestingly, this improvement does not appear at the first non-trivial order since there is no way to increase the number of independent connections at one node. In fact, generalized entanglement swappings remove such connections, so that only the nodes where no quantum operation is performed satisfy the inequality $\hat{\theta}(p) > \theta(p)$.

The previous analyses show that the multipartite strategy works well near the classical thresholds and near the ideal situation of perfect connections. We thus expect that our transformations give $\hat{P}(A \circ\!\!-\!\!\circ B) > P(A \circ\!\!-\!\!\circ B)$ for all values of entanglement in the links of the quantum networks. We performed Monte Carlo simulation for the eight lattices (using importance sampling via the perimeter method in the high-density region), attesting that $\hat{\theta}(p)$ is indeed strictly larger than $\theta(p)$ for $p \in (\hat{p}_c, 1)$. Numerical results are shown for three lattices in Fig. 2.13.

2.4 On optimal protocols

It is now legitimate to wonder about the optimality of the protocols based on multipartite entanglement percolation. First, it is not difficult to see that they cannot be optimal for every value S of initial entanglement in the links, at least for those lattices which admit a deterministic strategy. In fact, we have seen in Sec. 2.1.2 that a long-distance perfect Bell pair can be obtained with unit probability in the square lattice if S is larger than the threshold $S_c \approx 0.649$, which

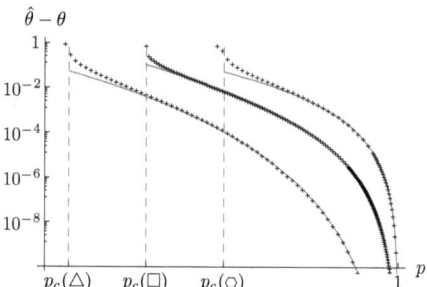

Figure 2.13: Numerical determination of $\hat{\theta} - \theta$, showing that multipartite percolation yields better long-distance entanglement for any initial entangled state $|\varphi\rangle$, with $p = 2\varphi_1$. We plot here the results for the triangle, square and hexagonal lattices (from left to right). Remark that the numerical data agree perfectly with the high-density expansions (gray lines) not only for p close to unity but also for much smaller values.

is possible in the multipartite protocol only if $S = 1$. However, remark that the fraction f of connected nodes with the deterministic strategy vanishes for infinitely large lattices. This fraction tends to the non-trivial value $f = 1/2$, nonetheless, if one considers the higher threshold $S^* \approx 0.684$, see Fig. 2.5. On the other hand, deterministic strategies fail completely if S lies between 0.5 and S_c, and therefore do not even reach the classical entanglement percolation threshold. Consequently, considering one strategy only for all values of S is not sufficient, but finding the optimum one for a given amount of entanglement in the links is a formidably difficult problem. In this respect, multipartite entanglement percolation is well-suited to generate long-distance quantum correlations regardless of the entanglement of the links, since it yields high connection probabilities and low thresholds at the same time.

Existence of a necessary amount of entanglement?

In what follows, we focus on the square lattice because it is the most "natural" one, but the arguments presented here apply to other lattices equally well. First, one easily sees that the threshold $S_c \approx 0.393$ given by the multipartite entanglement percolation strategy is not optimum. In fact, it can be slightly improved by considering an iterative measurement scheme. In the current protocol, one performs a generalized entanglement swapping between two partially-entangled pairs of qubits, see Eq. (2.14), and gets either a perfect GHZ state of three qubits or a weakly entangled one. The latter state has less entanglement than previously, with $\varphi'_1 = \varphi_1/(\varphi_0^2 + \varphi_1^2)$, and is distilled into the perfect state $|GHZ_3\rangle$ with probability $2\varphi'_1$, leading to an overall conversion probability of $2\varphi_1$. Instead of distilling the weakly entangled states, one can rather try to percolate entanglement within the (double-size and partially filled) square lattices they form. In fact, it is advantageous to create the state $|GHZ_5\rangle$ from two adjacent weakly entangled states of three qubits, rather than converting them separately. This procedure can be further iterated, creating quantum states that span $2^l + 1$ nodes at each level l of the iteration. However, the probability that a GHZ state is created at a level l is approximatively equal to $2\varphi_1^{(2^{l-1})}$, so that only the very first levels improve the threshold for $2\varphi_1 \approx S_c$. For instance, setting $l \leq 3$, one finds the new threshold

$$S_c = 0.375(4), \tag{2.19}$$

which gives the lowest amount of entanglement allowing the generation of Bell pairs over a large distance for all protocols known at the moment.

In order to prove the existence of a minimum threshold for propagating entanglement over infinite distances, one would be tempted to have the following reasoning (which is similar to the proof for bond percolation). Suppose that all paths Γ originating at one node are considered as being independent, that is, a failure on one path does not affect the others. Then, it is clear that the entanglement is confined to a finite region if the inequality

$$\sum_{N,\Gamma_N} S^{(N)} \leq \sum_N \mu_2^N S^{(N)} \leq \sum_N \mu_2^N C_{\max}^{(N)} = \sum_N \left(2\mu_2\sqrt{\varphi_0\varphi_1}\right)^N < \infty \tag{2.20}$$

holds for $N \to \infty$, where $\mu_2 \lesssim 2.679$ is the *self-avoiding walk connective constant* of the square lattice [PT00], and where the intermediate (in)equalities are derived in Sec. 1.3.1. This is the case if $\varphi_1 < \varphi_1^* \approx 3.61\%$, meaning that the minimum threshold would be lower bounded by the value $S^* \approx 7.22\%$. But it is not clear at all to which extent the paths can be treated independently, because their quantum nature may lead to constructive interferences. In conclusion, while any percolation protocol defines a sufficient condition in higher-dimensional systems, the question whether there exists or not a necessary amount of initial pure-state entanglement for generating long-distance entanglement is still open.

CHAPTER 3

Quantum complex networks

Complex networks describe a wide variety of systems in nature and society [BS03, Wat04, NBW06]. They model, among other things, chemical reactions in a cell, the spreading of diseases in populations, the predator-prey relationships between species, or the connections between routers and computers in the Internet; see, for example, [AB02] for a nice review. Due to the increasing computing power and the emergence of large databases on many real networks, they have attracted a lot of attention in the scientific community in the past few years. Many measures have been suggested to quantify some properties of their rich topology, but three concepts seem to occupy a prominent place: the small-world [Koc89], clustering [WS98] and scale-free [BA99] behaviors. The first approach to such networks was proposed by the Hungarian mathematicians Paul Erdős and Alfréd Rényi in the 1950's and 1960's. In a series of seminal papers [ER59, ER60, ER61], they introduced probabilistic methods in the theory of regular graphs, giving rise to the branch of *random graphs*. These graphs, which exhibit the important and universal small-world property,[1] opened the way to the study of complex networks. While many other models have been introduced over the years, random graphs are still widely used as they provide a reference for empirical studies in many fields.

In the introduction of this Thesis, we have seen that networks based on the laws of quantum physics, which allow perfectly secure communication, are expected to be developed in a near future. Since the actual communication networks, such as the Internet or phone-call networks, have a complex topology, it is natural to address the possibility that quantum networks will share this property too. Recently, in a somewhat different context, some links between complex networks and the quantum world have been made [JKBH08, CC09], and in this chapter we introduce a model of *quantum complex networks*, a new class of systems that exhibit some totally unexpected properties. Our model, described in Sec. 3.1, uses the random networks of Erdős and Rényi as a first case study and will certainly allow in the future the discussion of more complex architectures. Contrary to the other chapters of this Thesis, here we do not investigate the problem of long-distance entanglement, since this very concept does not make much sense in small-world networks. We thus turn to yet another property of random graphs, namely the appearance of subgraphs according to the connection probability (Sec. 3.1.1), and obtain a completely different classification of the behavior of quantum networks as compared to their classical counterpart (Sec. 3.3).

---

[1] If the number of random connections is large enough, the shortest path length l between any pair of nodes varies typically as the logarithm of the network size: $\langle l \rangle \sim \log(N)$.

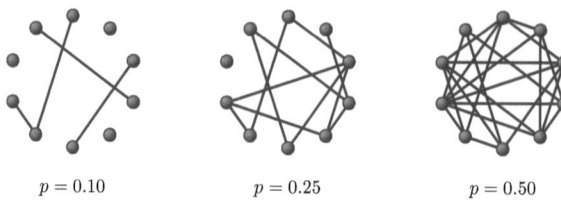

| $p = 0.10$ | $p = 0.25$ | $p = 0.50$ |

Figure 3.1: Evolution process of a random graph of size $N = 10$: starting from isolated nodes, we randomly add edges with increasing probability p, to eventually get the complete graph K_{10} for $p = 1$.

3.1 The model

In this section, we first recall how random graphs are generated and then propose an extension to the quantum world.

3.1.1 Random graphs

Let us now introduce the basics of random graph theory; the interested reader is referred to the original articles [ER59, ER60, ER61] or to Chap. III in [AB02] for a more detailed description of these graphs and a rigorous discussion of their properties. The theory of random graphs considers graphs in which each pair of nodes i and j are joined by a link with probability $p_{i,j}$. The simplest and most studied model is the one where this probability is independent of the nodes, with $p_{i,j} = p$, and the resulting graph is denoted $G_{N,p}$. The construction of these graphs can be considered as an evolution process: starting from N isolated nodes, random edges are successively added and the obtained graphs correspond to larger and larger connection probability, see Fig. 3.1. One of the main goals of random-graph theory is then to determine at which probability p a specific property $\mathcal{P}$ of a graph $G_{N,p}$ mostly arises, as N tends to infinity. Many properties of interest appear suddenly, *i.e.*, there exists a critical probability $p_c(N)$ such that almost every graph has the property $\mathcal{P}$ if $p \geq p_c(N)$ and fails to have it otherwise; such graphs are said to by *typical*. In our case, we are interested in the appearance of subgraphs (some specific patterns of links between the nodes) according to the connection probability p.

Subgraphs

We define a subgraph $F = (V, E)$ of $G_{N,p}$ as a collection of $n \leq N$ vertices V connected by l edges E. We restrict F to be connected and of finite size. One can ask the question: for which value of p is the subgraph to be found in a typical random graph? The answer was given in [Bol85], where the critical probability p_c for the appearance of F was proven to be

$$p_c(N) = c\, N^{-n/l}, \tag{3.1}$$

z	$-\infty$	-2	$-\frac{3}{2}$	$-\frac{4}{3}$	-1	$-\frac{2}{3}$
F	•	•—•	⋀	⋀	△□	⊠

Table 3.1: Some critical probabilities, according to Eq. (3.1), at which a subgraph F appears in random graphs of N nodes connected with probability $p \sim N^z$. For instance, cycles and trees of all orders appear at $z = -1$, whereas complete subgraphs (of order four or more) appear at a higher connection probability.

with c being independent of N. It is instructive to look at the appearance of subgraphs assuming that $p(N)$ scales as N^z, with $z \in (-\infty, 0]$ a tunable parameter: as z increases, more and more complex subgraphs emerge, see Tab. 3.1. In particular, only trivial connections appear in the regime $z = -2$.

Degree distribution

It is clear that the number of links attached to a node i, written m_i, follows a binomial distribution with parameters $N-1$ and p:

$$\text{prob}(m_i = m) = \frac{(N-1)!}{m!(N-1-m)!} p^m (1-p)^{N-m-1}. \tag{3.2}$$

It is known that the degree distribution of a random graph is well approximated by the distribution of independent nodes, which yields, for large N, the Poisson distribution

$$\text{prob}(m) \simeq e^{-\langle m \rangle} \frac{\langle m \rangle^m}{m!}, \tag{3.3}$$

with $\langle m \rangle = pN$. This degree distribution is a specific property of the random graphs and contrasts with the power law of the scale-free networks. Setting $p = cN^{-2}$, i.e., considering the first non-trivial random graphs, one therefore finds that there are in average $c^m N^{1-m}/m!$ nodes of degree m. Consequently, for sufficiently large c and for N tending to infinity, the number of nodes with $m = 1$ can be approximated by a Gaussian distribution of mean c and standard deviation $\sqrt{c}$, while no node with $m \geq 2$ appears in the network.

3.1.2 Erdős-Rényi networks in the quantum world

We consider now the natural extension of the previous scenario to a quantum context. For each pair of nodes, we replace the probability $p_{i,j}$ by a quantum state $\rho_{i,j}$ of two qubits, one at each node. Hence, every node possesses $N-1$ qubits which are pairwise entangled with the qubits of the other nodes, see Fig. 3.2a. As in the classical random graphs, we consider that pairs of particles are identically connected, with $\rho_{i,j} = \rho$. Furthermore, we restrict ourselves to the simplest case of pure states of qubits, that is, $\rho = |\varphi\rangle\langle\varphi|$, since it already leads to some very

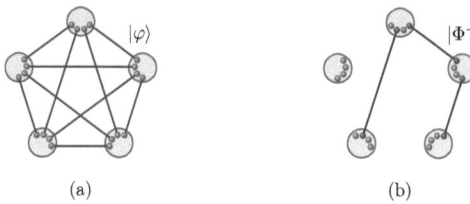

Figure 3.2: An example of a quantum random graph on five nodes. (a) Each node is in possession of four qubits which are entangled with the corresponding qubits at the other nodes. All the connections are given by the non-maximally entangled state $|\varphi\rangle$. (b) The quantum links can be converted into Bell pairs with probability $p = 2\varphi_1$ (here, $p = 0.25$). This strategy mimics the behavior of classical random graphs.

intriguing phenomena. Similarly to the previous chapters, we take these states to be

$$|\varphi\rangle \equiv \sqrt{1 - \frac{p}{2}}|00\rangle + \sqrt{\frac{p}{2}}|11\rangle, \qquad (3.4)$$

where $0 \leq p \leq 1$ measures the degree of entanglement of the links. As in the classical case, p scales with N and we write

$$|G_{N,p}\rangle \equiv \bigotimes_{i<j=1}^{N} |\varphi\rangle_{ij} \qquad (3.5)$$

the corresponding quantum random graph. Expanding in the computational basis all terms of this last expression, one notes that $|G_{N,p}\rangle$ is nothing but the coherent superposition of all possible simple graphs on N nodes, weighted by their number of links or "excitations" $|11\rangle$. For instance, drawing a line[2] for the states $|11\rangle$ and nothing for $|00\rangle$ (the qubits are not explicitly shown), the quantum state on three nodes reads:

$$|G_{3,p}\rangle = \sqrt{\varphi_0}^3 \left|\begin{smallmatrix}\bullet\\\bullet\;\bullet\end{smallmatrix}\right\rangle + \varphi_0\sqrt{\varphi_1}\left(\left|\begin{smallmatrix}\bullet\\\bullet\!\diagdown\!\bullet\end{smallmatrix}\right\rangle + \left|\begin{smallmatrix}\bullet\\\bullet\!-\!\bullet\end{smallmatrix}\right\rangle + \left|\begin{smallmatrix}\bullet\\\bullet\!\diagup\!\bullet\end{smallmatrix}\right\rangle\right) \\ + \sqrt{\varphi_0}\,\varphi_1\left(\left|\mathrel{\angle}\right\rangle + \left|\mathrel{\wedge}\right\rangle + \left|\mathrel{\searrow}\right\rangle\right) + \sqrt{\varphi_1}^3\left|\triangle\right\rangle, \quad (3.6)$$

with $\varphi_1 = p/2$ and, as usual, $\varphi_0 = 1 - \varphi_1$. The choice of these Schmidt coefficients becomes clear if one considers the "classical" strategy in which one tries to convert each link, independently and using LOCC only, into the Bell pair $|\Phi^+\rangle$. From Eq. (1.9), we know that the optimal probability of a successful conversion is p, and therefore the task of determining the type of maximally entangled states remaining after these conversions is mapped to the classical problem. In some sense, it corresponds to picking out a random but typical graph from the coherent superposition of all weighted graphs, but where a link designates now a Bell pair. Thus, we obtain the results of Tab. 3.1, and for example for $z \geq -2$, the probability to find a pair of nodes sharing a maximally entangled state is one, whereas that of having three nodes that share three maximally entangled

---

[2]Note that, depending on the context, a line between two nodes represents a separable state, a partially entangled one, or a Bell pair.

states is zero unless $z \geq -1$.

From now on, we set $p = 2cN^{-2}$, which is, as in the classical case, the first non-trivial regime for quantum random graphs. In fact, for $z < -2$ the overlap of $|G_{N,p}\rangle$ and the product state of all qubits in $|0\rangle$ approaches unity in the limit of infinite lattices:

$$|\langle G_{N,p}|0\ldots 0\rangle|^2 = \left(1 - \frac{p}{2}\right)^{\frac{N(N-1)}{2}} \simeq \exp\left(\frac{-N^{z+2}}{4}\right) \xrightarrow[N\to\infty]{z<-2} 1, \quad (3.7)$$

in which case no local quantum operation is able to create entanglement between the nodes.

3.2 Joint measurements help

Allowing strategies which entangle the qubits within the nodes offers new possibilities and brings powerful results. This is indeed a general statement in the context of quantum networks, as illustrated in the various chapters of this Thesis. In the current case, the constructions are based on the incomplete measurement $\mathcal{P}_M$, whose elements P_m are projectors onto the subspaces consisting of exactly m excitations $|1\rangle$ out of M qubits:

$$P_m \equiv \sum_{\pi_m} \pi_m |\underbrace{0\ldots 0}_{M-m}\underbrace{1\ldots 1}_{m}\rangle\langle 0\ldots 01\ldots 1| \pi_m^\dagger, \quad (3.8)$$

where π_m designates a permutation of the qubits. Applied on a node of the network, the measurement $\mathcal{P}_{N-1}$ counts the total number m of (separable) links $|11\rangle$ attached to it, without revealing their precise location, however. Remark that the value of the random outcome m is either 0 or 1 in the regime $z = -2$, since the probability to get larger values vanishes exponentially with N. Moreover, applied on all nodes of the network, the outcomes $m = 1$ follow the classical degree distribution: they approach the normal distribution $\mathcal{N}(\mu, \sigma^2)$ with $\mu = \sigma^2 = c$. In the rest of this section, we describe how the operators P_m can be used to extract some multipartite states that are pertinent in quantum information theory, such as the W and the GHZ states.

3.2.1 Creation of W states

Starting from $|G_{N,p}\rangle$ with $p = 2cN^{-2}$, we show in this section how to obtain the W state on n particles,

$$|W_n\rangle \equiv \frac{|10\ldots 0\rangle + |01\ldots 0\rangle + \ldots + |0\ldots 01\rangle}{\sqrt{n}}, \quad (3.9)$$

for any $n \ll N$. Restricting n to be much smaller than N is not strictly necessary, but the picture of the situation is clearer and the generalization to $n \lesssim N$ is straightforward. The structure of this state is pretty related to the one of the measurement $\mathcal{P}$, so that it seems to be a natural starting point for the construction of relevant multipartite quantum states. To that

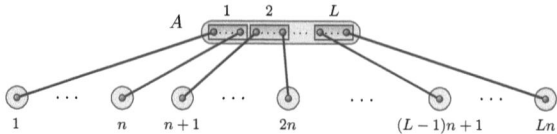

Figure 3.3: Construction of $|W_n\rangle$: one applies L times the measurement $\mathcal{P}_n$ on the qubits of a node A, which has $M = Ln$ neighbors and was previously projected onto the subspace consisting of one excitation $|1\rangle$. This randomly selects a group of n qubits sharing the excitation, which are thus in the state $|W_n\rangle$.

end, let us sequentially apply the measurement operators P_m on the nodes of the network until we get the outcome $m = 1$ for some node A. The fact that the outcomes m follow the normal distribution $\mathcal{N}(c, c)$ ensures that, for sufficiently large (but constant) c, the node A is found while $M \gg n$ unmeasured nodes are still present in the network. Without loss of generality, we assume that M is a multiple of n: $M = Ln$ for some integer L. In fact, we can always measure a small number of qubits of A without detecting the excitation $|1\rangle$. Then, we discard all links which are shared among the remaining M nodes, *i.e.*, we measure in the computational basis the qubits that are not connected to A. Finally, we perform L times the measurement $\mathcal{P}_n$ on the qubits of A, as depicted in Fig. 3.3. Exactly one measurement outputs P_1, and the state of the corresponding qubits reads:

$$|w_n\rangle = \frac{1}{\sqrt{n}} \left(|1\rangle |10\ldots 0\rangle + |2\rangle |01\ldots 0\rangle + \ldots + |n\rangle |00\ldots 1\rangle \right), \qquad (3.10)$$

where the first ket of each term belongs to A (and has been relabeled for convenience), while the other qubits are distributed among its n neighbors. At this point, let us introduce the notation

$$|\Phi_l^{k,d}\rangle \equiv \frac{1}{\sqrt{d}} \sum_{j=1}^{d} e^{\frac{2\pi i}{d} jk} |\underbrace{jj\ldots j}_{l}\rangle, \quad k \in \{1, \ldots, d\}, \qquad (3.11)$$

for the Fourier transform of $|\Phi_l^d\rangle \equiv |\Phi_l^{d,d}\rangle$, the GHZ state on l particles of dimension d. It is now a matter of fact to get $|W_n\rangle$ from $|w_n\rangle$: first, we measure A in the Fourier basis $\{|\Phi_1^{k,n}\rangle\}$, and second, given the outcome k, we get rid of the possibly introduced phases by applying the unitary

$$\begin{pmatrix} 1 & 0 \\ 0 & e^{-\frac{2\pi i}{n} jk} \end{pmatrix} \qquad (3.12)$$

on each of the remaining qubits, labeled by j.

3.2.2 Creation of GHZ states

As a second illustration of the advantage of joint actions on the qubits of the nodes, let us show how GHZ states of n qubits can be created in a quantum random graph. Contrary to the

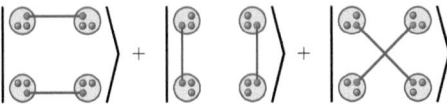

Figure 3.4: Graphical representation of $|K_c\rangle$. A link denotes, here, the separable state $|11\rangle$ while $|00\rangle$ is not drawn, and $|K_c\rangle$ is defined as the coherent superposition of all perfect matchings of the complete graph K_c. In this example, an appropriate labelling of the states leads to $|K_4\rangle \propto |1111\rangle + |2222\rangle + |3333\rangle$, which is a GHZ state of four qutrits.

previous example, n has to be finite, that is, independent of N, and without loss of generality we consider even n only. In fact, a GHZ state of odd size n can be obtained by measuring in the X basis one qubit of a GHZ state of size $n+1$. The construction proceeds in two steps. First, a highly-entangled multipartite state is created from the quantum network with probability approaching unity. Second, some qubits are measured in the computational basis, and a GHZ state is obtained if all measured qubits were in the state $|0\rangle$. The second step is therefore a probabilistic process, but as we will see, the probability of a successful construction can be arbitrarily amplified in the limit of infinite lattices.

Quantum perfect matchings in complete graphs

The first operation consists of performing the measurement $\mathcal{P}_{N-1}$ at all nodes of the quantum network $|G_{N,p}\rangle$. For $p = 2cN^{-2}$, we get the outcome $m = 0$ at nearly all nodes, and these nodes factor out. This is because the corresponding qubits are completely uncorrelated with the rest of the system (they are in the separable state $|0\rangle$). All other outcomes are $m = 1$ in the limit $N \to \infty$, and for large enough c we have seen that the number of such outcomes is distributed according to the normal function $\mathcal{N}(c,c)$. For creating GHZ states of size n, we need at least n outcomes $m = 1$, which is achieved with a probability exponentially close to unity by setting $c \gtrsim n + \sqrt{n}$. Suppose now that we get $c' > n$ outcomes $m = 1$, i.e., we have a quantum system that consists of c' nodes sharing $c'/2$ excitations $|11\rangle$; for clarity, let us simply write c instead of c' in what follows. Remark that there is always an even number of outcomes $m = 1$, so that $c/2$ is an integer. The remaining state is denoted by $|K_c\rangle$, and each node possesses exactly $c - 1$ orthogonal states in which $c - 2$ qubits are $|0\rangle$ and one is $|1\rangle$. Qubits that were originally entangled share the same state, and therefore this state can be visualized as the superposition of all perfect matchings of the complete graph K_c, see Fig. 3.4.

The size of $|K_c\rangle$ can then be decreased in a deterministic manner: one measures all qubits of a node i, and gets 0 for all outcomes except for a random one, say j. In fact, exactly one excited link is attached to i, and it equally points to all other nodes. Hence, the nodes i and j factor out, and the state is projected onto $|K_{c-2}\rangle$. One can sequentially repeat this procedure, which eventually produces the state $|K_n\rangle$.

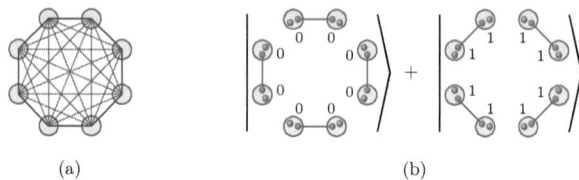

Figure 3.5: Creation of the state $|\text{GHZ}_n\rangle$ from $|K_n\rangle$. (a) Measurement pattern for $n=8$: all inner links (dashed thin lines) are measured in the Z basis (for clarity, the corresponding qubits are not explicitly drawn). (b) If all qubit outcomes are 0, then one gets the state $|\text{GHZ}_8\rangle$.

Construction of GHZ states and probability amplification

Let us show in this paragraph how $|K_n\rangle$ can be probabilistically transformed into a GHZ state of n qubits. Written in the computational basis, the state $|K_n\rangle$ has $(n-1)!!$ distinct terms,[3] which corresponds to the number of perfect matchings of K_n, while a GHZ state has only two of them. We thus have to "remove" the unwanted quantum correlations that are present in $|K_n\rangle$, which is done by measuring in the Z basis $n-3$ qubits at each node, as depicted in Fig. 3.5. The construction is successful if all outcomes are 0, so that the $n/2$ excitations $|11\rangle$ are still shared among the n nodes. This happens with probability $p_n = 2/(n-1)!!$, and in this case only two perfect matchings are possible in the underlying graph (a loop of length n). Therefore, the two qubits of each node are either in the state $|01\rangle$ or $|10\rangle$, and we project them onto the states $|0\rangle$ and $|1\rangle$, respectively, which leads to the desired GHZ state.

The GHZ state is successfully generated if we get some specific outcomes for the various measurements at the nodes, namely 0 for each qubit, and therefore it appears with non-unit probability p_n. But this is not a problem because p_n does not depends on N, so that the process can be arbitrarily amplified in the limit of infinite lattice size. The reason is that we can always subdivide the N nodes of the original network into L sets of N/L nodes, with L a constant much larger than $1/p_n$, and apply the construction of the GHZ state on each set. These sets are treated as independent if we initially discard all links connecting different sets, *i.e.*, if we measure the corresponding qubits (in any basis). Consequently, the GHZ states can be obtained with a probability that is arbitrary close to unity. Because of this probability amplification, we do not try to optimize the construction, which could, however, be essential for practical purposes since p_n decreases exponentially with n.

---

[3] The double factorial of a positive integer n is defined as $n!! \equiv n \cdot (n-2) \ldots 3 \cdot 1$ for odd n, and as $n \cdot (n-2) \ldots 4 \cdot 2$ for even n.

3.3 A complete collapse of the critical exponents

In the previous section, we have shown how to create some well-known multipartite entangled states, namely the W and the GHZ states of n qubits. These states do not have any classical counterpart, so let us now turn to the creation of *quantum subgraphs*. We define a quantum subgraph of a subgraph $F = (V, E)$ composed of n vertices and l edges as the state $|F\rangle$ consisting of l maximally entangled pairs $|\Phi^+\rangle$ shared among n nodes, according to E:

$$|F\rangle \equiv \bigotimes_{i=1}^{l} |\Phi^+\rangle_{E_i}. \tag{3.13}$$

In Sec. 3.3.1, we consider the "Λ" subgraph, which consists of two edges sharing one common node. In the classical random graphs, this subgraph only appears for $z \geq -3/2$, see Tab. 3.1, but in what follows we present a construction to get the corresponding quantum subgraph already in the regime $z = -2$. This will give us some insights into the design of larger quantum subgraphs, and in Sec. 3.3.2 we indeed present a general construction to obtain *any* quantum subgraph of finite size from the state $|G_{N,p}\rangle$ with $p \sim N^{-2}$. This is an unexpected result, as all critical exponents of the classical random graphs collapse onto the first non-trivial value $z = -2$ in the quantum context.

3.3.1 The Λ subgraph

Let us show how one can extract two Bell pairs on three nodes A, B and C, as depicted in Fig. 1.1a, in the regime $z = -2$. Explicitly, we want to create the state

$$|\Lambda\rangle_{ABC} \equiv \frac{(|00\rangle + |11\rangle)_{AB}}{\sqrt{2}} \otimes \frac{(|00\rangle + |11\rangle)_{BC}}{\sqrt{2}}$$
$$= \frac{1}{2}(|0000\rangle + |0011\rangle + |1100\rangle + |1111\rangle)_{ABBC}. \tag{3.14}$$

We start the construction by creating the state $|K_6\rangle$, as described in Sec. 3.2.2. Then, we do not remove all inner links of the underlying complete graph (by measuring their qubits in the Z basis) but only five of them, see Fig. 3.6. With probability $p_\Lambda = 4/15$ all outcomes are 0, which results in the (unnormalized) state $|111111\rangle + |122122\rangle + |333333\rangle + |444433\rangle$ on $A, B, \ldots, F$. Then, we remove the nodes D, E and F by a measurement in the Fourier basis $\{|\Phi_1^{k,4}\rangle\}$, and we correct the possibly introduced phases such that we get the state

$$\frac{1}{2}(|111\rangle + |122\rangle + |333\rangle + |444\rangle)_{ABC}. \tag{3.15}$$

The last step of the construction consists of relabeling the states of B, with $|1\rangle \mapsto |00\rangle$, $|2\rangle \mapsto |01\rangle$, $|3\rangle \mapsto |10\rangle$ and $|4\rangle \mapsto |11\rangle$, and to project A and C onto the subspace of one qubit

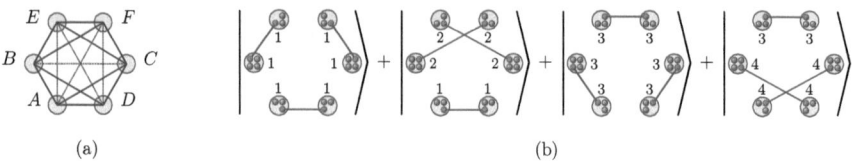

Figure 3.6: Construction of $|\Lambda\rangle$ from $|K_6\rangle$. (a) The five dashed links are measured, while the other qubits are left unaltered. (b) If all outcomes are 0, we get the quantum superposition of the four possible perfect matchings of the remaining graph. The states at the nodes are labeled according to the qubit which is in the state $|1\rangle$.

by applying the operators

$$P_{A\pm} = |0\rangle\frac{\langle 1| \pm \langle 2|}{\sqrt{2}} + |1\rangle\frac{\langle 3| \pm \langle 4|}{\sqrt{2}},$$
$$P_{C\pm} = |0\rangle\frac{\langle 1| \pm \langle 3|}{\sqrt{2}} + |1\rangle\frac{\langle 2| \pm \langle 4|}{\sqrt{2}}, \quad (3.16)$$

which satisfy the completeness relations $P_{A+}^\dagger P_{A+} + P_{A-}^\dagger P_{A-} = \mathbb{1}_4$ and $P_{C+}^\dagger P_{C+} + P_{C-}^\dagger P_{C-} = \mathbb{1}_4$. Depending on the outcomes, some local unitaries are applied on the state, such that it becomes the quantum subgraph $|\Lambda\rangle$ given in Eq. (3.14). Since the probability p_Λ does not depend on N, the construction can be repeated as many times as we want for N tending to infinity, so that $|\Lambda\rangle$ is generated with a probability arbitrarily close to unity. Therefore, the classical critical exponent $z = -3/2$ is shifted to $z = -2$ in the quantum setting.

3.3.2 General subgraphs

We now turn to the main result of this chapter:

Proposition 3.1 (Collapse of the critical exponents)

Let $|G_{N,p}\rangle$ a quantum random graph with $p = 2cN^{-2}$. For N tending to infinity, one can obtain, with probability approaching unity, the quantum state $|F\rangle$ with the structure of any finite subgraph $F = (V, E)$ composed of n vertices and l edges, as defined in Eq. (3.13).

This result implies that in the quantum case the structure of Tab. 1 completely changes, as all subgraphs already appear at $z = -2$. More precisely, for any subgraph F, there exists a quantum random graph with $p \sim N^{-2}$ where F appears. Note that the only restriction on F is that both n and l are independent of N. In particular, F does not need to be simple, as in the classical case.

PROOF The construction of $|F\rangle$ from a typical quantum random graph mainly follows the creation of the Λ subgraph: one tries to obtain a GHZ state of size n and of sufficiently large dimension d, as it was (nearly) the case in Eq. (3.15), and then performs some projections at the nodes in order to reveal the structure of F. Without loss of generality, we consider connected subgraphs only, such that $l \geq n-1$. Let us now detail the construction, which we split in three steps.

First step. We create the state $|K_c\rangle$ with $c = D+n$, $D = d^2$ and $d = 2^l$: n nodes will be kept to build the final state, while D additional nodes are needed to establish the desired quantum correlations. This state is obtained with unit probability by tuning the prefactor in p (and not the exponent $z = -2$), as described in Sec. 3.2.2. Then, we remove all connections shared between n nodes of $|K_c\rangle$, i.e., we measure the $n(n-1)/2$ concerned links. The operation is successful if all outcomes are 0. In that way, we build a state that is the coherent superposition of all perfect matchings of the graph join of K_D and the empty graph $\overline{K}_n$. Counting the number of perfect matchings of K_c and $K_D + \overline{K}_n$, one finds that the success probability of this operation is

$$\text{prob}\left(|K_c\rangle \mapsto |K_D + \overline{K}_n\rangle\right) = \frac{D!}{(D-n)!} \cdot \frac{(D-n-1)!!}{(D+n-1)!!}, \quad (3.17)$$

which approaches one for large n (and consequently for large D since $D = 2^{2l} \geq n$ for $l+1 \geq n \geq 2$). We further measure the D nodes of K_D, but this time in the Fourier basis $\{|\Phi_1^{k,c-1}\rangle\}$ in order not to reveal where the links $|11\rangle$ lie. Since the nodes can communicate their measurement outcomes, we can correct the possibly introduced phases, and the resulting state reads:

$$|\varphi_n^D\rangle = \sqrt{\frac{(D-n)!}{D!}} \sum_{i_1 \neq i_2 \ldots \neq i_n = 1}^{D} |i_1 i_2 \ldots i_n\rangle. \quad (3.18)$$

Third step. We now want to transform $|\varphi_n^D\rangle$ into a GHZ state of n particles of dimension $d = \sqrt{D}$. To that end, and because it is not convenient to deal with sums whose indices are subject to constraints, we develop Eq. (3.18) in order to let the sums freely run from 1 to D. For example, the state on three nodes A, B, and C is expressed as (up to a normalization factor):

$$|\varphi_3^D\rangle_{ABC} \propto \sum_{i,j,k=1}^{D} |ijk\rangle_{ABC} - \sum_{i,j=1}^{D} \left(|iij\rangle + |iji\rangle + |jii\rangle\right)_{ABC} + 2\sum_{i=1}^{D} |iii\rangle_{ABC}$$
$$= \sqrt{D}^3 |\Phi_1^D\rangle_A \otimes |\Phi_1^D\rangle_B \otimes |\Phi_1^D\rangle_C - D\left(|\Phi_2^D\rangle_{AB} \otimes |\Phi_1^D\rangle_C \right.$$
$$\left. + |\Phi_2^D\rangle_{AC} \otimes |\Phi_1^D\rangle_B + |\Phi_2^D\rangle_{BC} \otimes |\Phi_1^D\rangle_A\right) + 2\sqrt{D}\, |\Phi_3^D\rangle_{ABC}.$$

More generally, this leads to a weighted and symmetric superposition of states of the form $\otimes_{i=1}^r |\Phi_{\lambda_i}^D\rangle$ for all partitions[4] $(\lambda_1, \lambda_2, \ldots, \lambda_r)$ of n. We want to remove all terms of this sum but the last one, which will allow us to obtain $|F\rangle$. To that purpose, we use the mathematical identity

$$|\Phi_m^D\rangle = |\Phi_m^{d^2}\rangle = |\Phi_m^d\rangle^{\otimes 2}, \quad (3.19)$$

---

[4]A partition of a positive integer is a way of writing it as a sum of positive integers. For example, the partitions of 4 are (4), (3,1), (2,2), (2,1,1) and (1,1,1,1).

which holds for all m and d: we split each node into two subsystems of dimension d and measure one of them in the Fourier basis. This operation is successful if the outcomes are $k = 1$ at all nodes but the last one, which should be $k = d - n + 1$. In fact, to see what happens to a state $|\Phi_m^d\rangle$, we note that

$$\langle \Phi_1^{k,d} | \Phi_m^{j,d} \rangle = \begin{cases} \frac{1}{\sqrt{d}} |\Phi_{m-1}^{j-k,d}\rangle & \text{if } m > 1, \\ \delta_{j,k} & \text{otherwise.} \end{cases} \quad (3.20)$$

Therefore, by sequentially measuring $\langle \Phi_1^{1,d} |$, a state $|\Phi_m^d\rangle$ shared among any $m < n$ nodes transforms as

$$|\Phi_m^d\rangle \equiv |\Phi_m^{d,d}\rangle \mapsto |\Phi_{m-1}^{d-1,d}\rangle \mapsto \ldots \mapsto |\Phi_1^{d-m+1,d}\rangle \mapsto 0$$

as long as $d \neq m$. But this is always the case because we consider connected subgraphs, so that $d \geq 2^l \geq 2^{n-1} \geq n > m$. Hence, all terms $|\Phi_m^D\rangle = |\Phi_m^d\rangle^{\otimes 2}$ with $m < n$ vanish while measuring the first $n - 1$ nodes, which results in the state $|\Phi_n^d\rangle \otimes |\Phi_1^{d-n+1}\rangle$. The state $|\Phi_1^{d-n+1}\rangle$ is finally removed by the last Fourier measurement, so that we are left with the GHZ state $|\Phi_n^d\rangle$.

Fourth step. The last step consists of transforming $|\Phi_n^d\rangle$ into $|F\rangle$. The procedure is completely analogue to the last part of the construction of the Λ subgraph, see Eqs. (3.14) to (3.16). First, we expand and write explicitly all the terms of $|F\rangle$ in the computational basis, and we group the qubits according to the connections E. This leads to a sum of product states of the form $\otimes_{j=1}^{n} |\varphi_{i,j}\rangle$ with $i = 1, \ldots, 2^l$ and $j = 1, \ldots, n$. Since we have chosen $d = 2^l$, we can apply the measurement element $\sum_{i=1}^{d} |\varphi_{i,j}\rangle\langle i|$ on each node j of $|\Phi_n^d\rangle$, which achieves the desired transformation and therefore concludes the proof. □

Part II

Mixed states

The truth is rarely pure and never simple.

— Oscar Wilde

CHAPTER 4

Towards noisy quantum networks

In the first part of this Thesis, we have shown how pure-state entanglement can be manipulated in noiseless networks to achieve tasks that are impossible in one-dimensional settings (long-distance entanglement generation, Chap. 2) or to demonstrate surprising effects that are absent in the corresponding classical model (sudden appearance of subgraphs in complex networks, Chap. 3). This brings deep insights into the interplay of the network geometry and the quantum operations at the nodes, emphasizing the predominant role of multipartite entangled states. In fact, both the connectivity of the vertices and the spread of entanglement over the nodes help in these tasks. The aim of this chapter is to determine to which extent the results collected so far apply to the more realistic scenario of noisy networks, and, eventually, to introduce some refined techniques to tackle systems undergoing random errors.

At this point, let us briefly describe a possible setup for an implementation of a quantum network, where atoms store the quantum information, and thus represent qubits, while photons are used to create remote entanglement. This is currently the most promising scenario for the realisation of quantum networks [Kim08]. In particular, we consider the case where *continuous-variable* entanglement contained in a two-mode squeezed light is transformed into *discrete* entanglement between two atoms trapped in distant high-quality cavities [KC04], see Fig. 4.1. Assuming perfect operations, one can drive the system so that its steady state reads:

$$|\varphi\rangle_{AB} = \sqrt{\frac{n+1}{2n+1}} |gg\rangle_{AB} + \sqrt{\frac{n}{2n+1}} |ee\rangle_{AB}, \qquad (4.1)$$

with $|g\rangle$ and $|e\rangle$ denoting the ground and excited atomic states, respectively, and where $n \gtrsim 1$ is a squeezing parameter. This entangled state of two qubits is equivalent to the pure state considered so far, but we want now to include in the description of the links of the network some of the imperfections that are present in any real system.

If one source of errors is preponderant in the network, then the links are well approximated by rank-two mixed states, *i.e.*, mixtures of two two-qubit pure states (Sec. 4.1). In this case, the mathematical properties of the noisy connections are not "too far" from those of the pure states, so that most results of Part I still apply. For instance, entanglement percolation is possible if the connections are subject to *amplitude damping* [BDJ09], and multipartite measurements can boost this process, see Sec. 4.1.1. The description of the links by amplitude-damping channels is valid if the main source of noise is a loss of energy from the quantum system (as a spontaneous emission of one photon from the atoms, or the scattering of the photons in the connections, see Sec. 8.3.5 in [NC00]), but many other scenarios can be imagined. For example, the light source in Fig. 4.1 may fail to emit any squeezed light with a non-negligible probability. This is an important but nevertheless not too severe source of errors, since it does not alter, for instance,

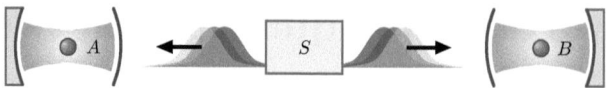

Figure 4.1: Two cavities A and B are simultaneously driven by a common source S of squeezed light, and the two distant atoms get entangled in the steady state of the system.

the properties of quantum complex networks (Sec. 4.1.2).

In a more general setting, errors affect all components of the four-by-four density matrices describing the links of the network, such that one has to deal with mixed states of full rank (Sec. 4.2). It is particularly important to use this description while studying the problem of long-distance entanglement generation, since locally negligible errors may add up to a complete loss of non-local quantum correlations. In Sec. 4.2.1, we review the basic properties of full-rank mixed states, and then we briefly describe the so-called *quantum-repeater* strategy (Sec. 4.2.2). In the last two chapters of this Thesis, we then show how the high connectivity of a general quantum network greatly improves the efficiency of these one-dimensional protocols.

4.1 Rank-two mixed states

In this section, we consider quantum networks that are exposed to one type of errors only, *i.e.*, the connections are described by a probabilistic mixture of two pure states. As for pure-state networks, we suppose that the quantum operations are not affected by the noise, so that they can be applied perfectly at the nodes.

4.1.1 From pure to mixed states and vice versa

As a general but unrigorous observation, rank-two mixed states behave quite classically, in the sense that one has to deal with one type of errors only. For example, one may face (classical) bit-flip errors while the (quantum) coherence of the state is undamaged. This observation is made more concrete by the following statements: a two-qubit pure state can be *twirled* to a mixed state of rank two [VW01], and some rank-two mixed states can be purified into Bell pairs [Jan02], starting from a finite number of copies.[1] In that respect, most ideas of Part I can be used here, modulo, of course, some adaptations of the formulas according to the setting under consideration.

---

[1] For two-qubit states of rank three or four, this is possible in the asymptotic limit only, that is, if one possesses an infinite number of copies.

Transforming weak pure-state entanglement into bit-flip errors

Before turning to mixed-state networks, let us note that the *twirl operation* [VW01], which transforms the state $|\varphi\rangle = \sqrt{\varphi_0}|00\rangle + \sqrt{\varphi_1}|11\rangle$ into the mixture

$$\rho = \frac{(\sqrt{\varphi_0} + \sqrt{\varphi_1})^2}{2}|\Phi^+\rangle\langle\Phi^+| + \frac{(\sqrt{\varphi_0} - \sqrt{\varphi_1})^2}{2}|\Psi^+\rangle\langle\Psi^+|, \qquad (4.2)$$

can be used to generated long-distance entanglement in pure-state lattices. In fact, we will see in Sec. 5.1 that useful quantum correlations between distant stations are maintained if the bit-flip error rate $\varepsilon_b \equiv (\sqrt{\varphi_0} - \sqrt{\varphi_1})^2/2$ does not exceed 10.94%. This offers an alternative to the percolation strategies, but it is a remarkable coincidence that both protocols lead to the same entanglement threshold. In fact, setting $\varepsilon_b^* = 10.94$, one finds that the critical amount of entanglement using mixed states is

$$S^* = 2\varphi_1^* \approx 0.3757, \qquad (4.3)$$

which is equal (up to numerical errors) to the value given in Eq. (2.19).

Percolation with amplitude-damping channels

It is a well-known result that two mixed states of the form

$$\rho(\varphi,\gamma) \equiv (1-\gamma)|\varphi\rangle\langle\varphi| + \gamma|01\rangle\langle01| \qquad (4.4)$$

shared between two parties, which describes the result of an amplitude-damping channel with energy dissipation parameter $0 < \gamma < 1$ (see Sec. 8.3.5 in [NC00]), can be purified into one Bell pair with a strictly positive probability of success and using LOCC only [Jan02]. In their recent paper [BDJ09], Broadfoot *et al.* nicely showed that this operation leads to mixed-state entanglement percolation in certain quantum networks. In what follows, we propose to incorporate the multipartite measurements described in Sec. 2.3 into their protocol, yielding lower and therefore advantageous entanglement thresholds. To that end, let us consider a triangular lattice, where each bond consists of the two mixed states $\rho(\Phi^+,\gamma)$ and $\rho(\varphi,\gamma)$. This is a very specific example, but the multipartite strategy clearly applies to more general situations. The protocol used in [BDJ09] first tries to purify every double bond, which results in the pure state $|\varphi\rangle$ with probability $(1-\gamma)^2/2$, and then it applies classical entanglement percolation. For instance, setting $\gamma = 1\%$, long-distance entanglement is possible if $S(\varphi)$ is larger than

$$S_c = \frac{2p_c^\triangle}{(1-\gamma)^2} \approx 0.7087. \qquad (4.5)$$

Alternatively, one could apply a generalized entanglement swapping on the bonds that are successfully purified, according to the multipartite measurement pattern for the triangular lattice, see Fig. 2.12. One subtlety arises here, however, since the lattice is not perfect but only partially filled with bonds. For simplicity, and in the very same spirit as the last part of Sec. 2.4, we do not adapt the measurement pattern to the random lattice but rather perform a GHZ measurement only if the two corresponding bonds are present. In that case, a numerical estimation of the new threshold yields

$$\hat{S}_c = 0.684(1), \tag{4.6}$$

which clearly beats the previous strategy.

4.1.2 Quantum complex networks

Let us reconsider the quantum complex networks of Chap. 3 in a slightly more realistic situation, where some errors are introduced into the system. For instance, the quantum connections may be deteriorated by a source of light that fails to emit squeezed states with some probability γ, see Fig. 4.1. Equivalently, this imperfect source produces the vacuum state $|00\rangle$ with probability γ, so that the connections of the quantum network are in the mixture

$$\rho = (1 - \gamma) |\varphi\rangle\langle\varphi| + \gamma |00\rangle\langle00|. \tag{4.7}$$

In what follows, we show that the construction of quantum subgraphs described in Sec. 3.3.2 is robust against this kind of noise in the connections of the network. In the first step of the proof, the measurements $\mathcal{P}_{N-1}$ are not affected by the imperfect source since no extra excitation $|1\rangle$ is created in the network: only the number of outcomes $m = 1$ slightly decreases from c to $(1 - \gamma)c$. But as already discussed, the connection probability p can be increased to still get with certainty c outcomes $m = 1$. Then, setting $c = 4$ for instance, the remaining nodes are in a mixture of $|K_4\rangle$ and some completely separable states:

$$\rho_{K4} = x |K_4\rangle\langle K_4| + \frac{1-x}{3} \sum_{i=1}^{3} |iiii\rangle\langle iiii|, \tag{4.8}$$

with $x = (1 - \gamma)^2$ for very large lattice size N. Despite the presence of separable states, ρ_{K4} is useful for quantum information tasks since it is distillable for all $\gamma < 1$, i.e., the coefficient x can be brought arbitrarily close to unity if one possesses a large number of such copies. However, in the regime $z = -2$, it is impossible to get several copies of ρ_{K4} on the *same* four nodes. But this is not a problem since, alternatively, one can repeat the construction $1/x$ times so that any use of $|K_4\rangle$ is still achieved with high probability. More generally, the state ρ_{Kc} is a mixture of the desired state $|K_c\rangle$ and some partially separable states, with x equal to $(1-\gamma)^{c/2}$. This structure is maintained throughout the construction of the quantum subgraphs (steps two to four of the proof), so that with a strictly positive probability x we create any quantum subgraph $|F\rangle$ in the regime $p \sim N^{-2}$.

4.2 Full-rank mixed states

From now on, we consider that the quantum connections between two neighboring qubits of the networks are completely general, that is, they are described by any valid density operator ρ. Such a four-by-four matrix can be expanded in the Bell basis, and its *fidelity*

$$F \equiv \langle \Phi^+ | \rho | \Phi^+ \rangle \tag{4.9}$$

describes how close ρ is to the ideal connection $|\Phi^+\rangle$. This state can always be *depolarized* to a standard form by applying random local unitaries on the two qubits, which greatly simplifies the calculations [BBP+96]. The resulting state, which is called a Werner state [Wer89], is diagonal in the Bell basis:

$$W_F \equiv F |\Phi^+\rangle\langle\Phi^+| + \frac{1-F}{3} \left(|\Psi^+\rangle\langle\Psi^+| + |\Phi^-\rangle\langle\Phi^-| + |\Psi^-\rangle\langle\Psi^-| \right),$$

and is entangled whenever $F > 1/2$ [BDSW96]. Note that the fidelity F is not altered by the depolarization procedure, and for convenience we rewrite this state as

$$W_x \equiv x |\Phi^+\rangle\langle\Phi^+| + \frac{1-x}{4} \mathbb{1}_4, \tag{4.10}$$

with $x = (4F-1)/3$. The latter definition emphasizes the fact that a Werner state is a mixture of a perfect quantum connection and a completely separable state.

Most concepts and results of this section are already well established, but their introduction here has two main purposes. First, it presents in a compact way the various quantum manipulations that have to be assimilated before turning to the last two chapters of this Thesis. This is done in Sec. 4.2.1. Second, it gives a flavor of what kind of entanglement can be extracted from the quantum connections. In particular, since LOCC transformations cannot purify or distill a finite number of such connections in a perfect manner, it can be anticipated that entanglement-percolation protocols (at least in their current form) will have to be abandoned for new strategies. In the last part of this chapter (Sec. 4.2.2), we finally review the so-called *quantum-repeater* protocols, which allow a rather efficient generation of long-distance entanglement in one-dimensional systems and traditionally serve as a benchmark for any new protocol.

4.2.1 Elementary operations on Werner states

In this subsection, we describe two basic quantum operations that allow one to manipulate the entanglement present in the Werner states, namely the entanglement swapping and the (bipartite) purification procedure. Then, we show how a Werner state is brought into yet another standard form (independent bit-flip and phase errors), which is the state that is considered in

Entanglement swapping and purification

Since the Werner state is a (classical) mixture of a Bell pair and a completely separable state, it trivially follows that the entanglement swapping described in Sec. 1.1.1 is optimally performed in any basis of maximally-entangled states. The state that is created by entanglement swapping between the two extremities of a chain of N Werner states W_x is thus given by

$$W_x^{(N)} = x^N |\Phi^+\rangle\langle\Phi^+| + \frac{1-x^N}{4} \mathbb{1}_4, \qquad (4.11)$$

so that the fidelity decreases exponentially with N. As expected, we face the same problem as with the pure states while trying to generate long-distance entanglement, and since the former was solved by means of distillation procedures, it is natural to study these methods in the mixed-state scenario. A lot of effort was furnished in this direction (see [DB07] for a review on entanglement purification), but to our purpose it is sufficient to notice that:

(i) At least two copies of a Werner state are needed to get, by LOCC and with finite probability, a state of higher fidelity [LMP98].

(ii) Perfect Bell pairs can be distilled from N Werner states in the limit $N \to \infty$ only [Ken98].

As an example of purification procedure, Bennett et al. presented a protocol to concentrate the entanglement of two states W_x and $W_{x'}$ [BBP+96], leading to a state $W_{x''}$ with

$$x'' = \frac{x + x' + 4\,xx'}{3 + 3\,xx'}. \qquad (4.12)$$

The resulting state is closer to the target state $|\Phi^+\rangle$ if both W_x and $W_{x'}$ are entangled (that is, if $x, x' > 1/3$) and if $x > x' > 2x/(1 + 4x - 3x^2)$. This procedure can be iterated, leading to a non-zero distillation rate, but improved schemes exist, see [DEJ+96] for example.

From Werner states to independent bit-flip and phase errors

Let us denote a Werner state W_F by its components in the Bell basis $\mathcal{B} = (\Phi^+, \Psi^+, \Phi^-, \Psi^-)$, so that $W_F = \left(F, \frac{1-F}{3}, \frac{1-F}{3}, \frac{1-F}{3}\right)_\mathcal{B}$. We want to find a sequence of separable operations on W_F in order to get the mixed state

$$\rho(\varepsilon_b, \varepsilon_p) \equiv \Big((1-\varepsilon_b)(1-\varepsilon_p),\ \varepsilon_b(1-\varepsilon_p),\ (1-\varepsilon_b)\varepsilon_p,\ \varepsilon_b\varepsilon_p\Big)_\mathcal{B}, \qquad (4.13)$$

which can be thought of as the result of sending one qubit of the Bell pair $|\Phi^+\rangle$ through a quantum channel that randomly introduces bit-flip or phase errors with independent probabilities ε_b

(a) 1st level   (b) 2nd level

Figure 4.2: The nested purification used by the quantum repeaters. (a) Elementary connections are continuously generated between neighboring stations, and an entanglement swapping is performed at every second node. The resulting states are weaklier entangled, but they are repeatedly purified by *entanglement pumping*. (b) Once the states have a sufficiently large fidelity, the procedure is iterated at a higher level, which eventually creates an entangled pair of qubits between the two extremities of the chain.

and ε_p, respectively. To that end, we first define the two unitaries

$$H = \frac{1}{\sqrt{2}} \begin{pmatrix} 1 & 1 \\ 1 & -1 \end{pmatrix} \quad \text{and} \quad H' = \frac{1}{\sqrt{2}} \begin{pmatrix} i & 1 \\ 1 & i \end{pmatrix}, \tag{4.14}$$

and we verify that the operation $H \otimes H$ applied on a Bell-diagonal state switches its coefficients Φ^- and Ψ^+, while the coefficients Φ^+ and Ψ^- are unaltered. A similar result holds for $H' \otimes H'$, which only switches the components Φ^+ and Ψ^+. Suppose now that an entangled pair W_F, with $F = 1 - 3\varepsilon_b\varepsilon_p$, has been created between two neighboring nodes. This already sets the coefficient Ψ^- to the desired value $\varepsilon_b\varepsilon_p$. Then, apply $H' \otimes H'$ with probability $p = (\varepsilon_b + \varepsilon_p - 4\varepsilon_b\varepsilon_p)/(1 - 4\varepsilon_b\varepsilon_p)$ and $\mathbb{1}_2 \otimes \mathbb{1}_2$ with probability $1 - p$. This sets the fidelity of the resulting state to its final value $(1 - \varepsilon_b)(1 - \varepsilon_p)$, while the second and third components read $\varepsilon_b + \varepsilon_p - 3\varepsilon_b\varepsilon_p$ and $\varepsilon_b\varepsilon_p$, respectively. Finally, repeat the operation by applying $H \otimes H$ with probability $p = \varepsilon_p(1 - 2\varepsilon_b)/(\varepsilon_b + \varepsilon_p - 4\varepsilon_b\varepsilon_p)$, which yields the desired result.

4.2.2 Quantum repeaters

The *quantum repeaters* decribed in [BDCZ98, DBCZ99] offered a first solution to the problem of quantum communication between distant parties.[2] In this scheme, the time needed for the generation of entanglement between the two extremities of a one-dimensional lattice of size N scales polynomially with N, whereas only a logarithmic number of qubits is required per station. The underlying idea of the quantum repeaters is to use a *nested purification protocol*, which intersperses entanglement swappings and purification steps, see Fig. 4.2.

Though being very promising, the quantum repeaters raise several technical problems, such as the difficulty to manipulate many qubits per station, or maybe more fundamentally, the need for reliable quantum memories [HKBD07]. The former is surmounted in [CTSL05], where a constant number of qubits is required at each station. Alternatively, a scheme involving

---

[2]Note that an anterior scheme based on concatenated quantum codes was proposed in [KL96]. However, even if the physical resources scale only polynomially with the length of the channel in that case, the large number of qubits at each station or the high precision with which the various quantum operations have to be realized may be out of reach for any practical implementation.

optical instruments only, such as laser manipulation, beam splitters, or single-photon detectors, was proposed in [DLCZ01]. Since then, various protocols improving the rate of long-distance quantum communication in one-dimensional networks have been developed [CZC$^+$07, JTL07, SdA$^+$07, SSZ$^+$08, JTN$^+$09]; see [SSdRG09] for a review on this topic. However, either their running time scales polynomially with the distance, or they are based on rather complicated quantum error-correcting codes involving many qubits per station.

4.2.3 Lower bound for long-range entanglement

The quantum repeater protocols were created for one-dimensional systems, and the purpose of the next two chapters is to design new and efficient schemes for distributing entanglement in noisy quantum networks of higher dimension. Before turning to this problem, however, let us first provide a bound on the fidelity of the connections under which no long-range entanglement can be generated.

In Chap. 2, we have seen that ideas from percolation theory allow one to set an *upper* bound on the minimum amount of pure-state entanglement required to connect two infinitely distant nodes, whereas no non-trivial lower bound is known at the moment (Sec. 2.4). Quite ironically, the situation turns out to be the opposite for mixed-state networks: a *lower* bound is easily derived from percolation considerations, whereas an upper bound is found for three-dimensional networks only (Chap. 6). In fact, suppose that the connections of the network are given by the Werner state $W_p = p|\Phi^+\rangle\langle\Phi^+| + (1-p)\mathbb{1}_4/4$, with p smaller than the (classical) threshold p_c for bond percolation in the corresponding lattice. The quantum state describing the whole system is then a mixture of lattices whose bonds are either perfect Bell pairs or completely separable states, but in the limit of infinite size, *none* of these lattices possesses a giant cluster of Bell pairs. This threshold leads to a lower bound, since, by definition, no local quantum operation can create entanglement out of separable states. In the square lattice, for instance, genuine quantum correlations cannot be generated over arbitrarily large distances if $p < p_c^\square = 1/2$, even though all connections are entangled in the range $p \in (\frac{1}{3}, \frac{1}{2})$. Remark that this argument does not provide any interesting lower bound for the cubic lattice, since its percolation threshold $p_c \approx 0.2488$ [LZ98] is smaller than one third.

CHAPTER 5

One-shot entanglement generation over large distance in square lattices

The aim of this chapter is to establish a new theoretical scheme for generating long-distance entanglement in quantum networks, without relying on efficient quantum memories [HKBD07] or on rather complicated concatenated quantum error-correcting codes [JTN+09]. We consider a $N \times N$ square lattice, where neighboring nodes are connected by noisy quantum channels, and we show that a partially entangled pair of qubits can be created between any two stations if the effective probability for the quantum errors lies below a certain threshold. To that purpose, we propose to combine ideas of quantum correcting codes [KL96] with error-recovery techniques developed in the context of topological quantum memory [DKLP02]. Our protocol is a "one-shot" process (the elementary entangled pairs are used only once) involving one-way classical communication only, and therefore the qubits have to be preserved from decoherence for a time that does not depend on the number of connections between the two desired stations. Furthermore, the overhead of local resources increases only logarithmically with N while the tolerable error probability for the various quantum operations is of the order of the percent for any realistic network size, making our proposal favorable for long-distance quantum communication.

In Sec. 5.1, we start by considering that only bit-flip errors randomly occur in the network. We show that most such defects can be suppressed if their occurrence rate does not exceed 11%. To this end, we construct a giant GHZ state that is distributed among all stations (Sec. 5.1.1), and the high connectivity of the network allows us to compute the parity of some pairs of its qubits, that is, to gain information on the location of the bit-flip errors,[1] see Sec. 5.1.2. Then, we show that a somewhat "classical" encoding of the qubits prevents the phase errors from proliferating in the network, which therefore preserves the coherence of the final quantum state (Sec. 5.2). Finally, we emphasize the fault-tolerant aspect of our protocol, relating the general model of errors that we use to some concrete physical quantities, such as the decoherence rate of the individual systems, or the loss of photons while generating the short-distance entangled pairs, see Sec. 5.2.2.

5.1 Network with bit-flip errors only

The quantum network we consider throughout this chapter consists of a $N \times N$ square lattice, where nodes represent the stations and edges the quantum channels, see Fig. 5.1. Quantum states can be transmitted through these channels, and entanglement, *i.e.*, short-distance Bell

---

[1]Since a GHZ state is the coherent superposition of the states $|00\ldots0\rangle$ and $|11\ldots1\rangle$, the parity of any two qubits is even in a noiseless scenario. An odd parity thus indicates that at least one bit-flip occurred in the network.

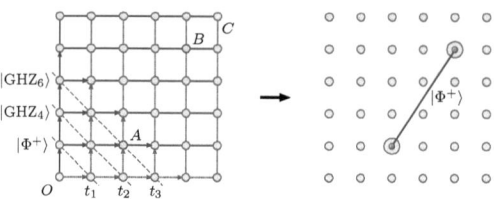

Figure 5.1: The links of a square lattice are used to teleport a GHZ state from the node O to the node C. Two stations A and B coherently keep one qubit of the propagating GHZ state, generating the long-distance Bell pair $|\Phi^+\rangle$.

states, can be created between neighboring stations. We would like to use these resources to generate a long-distance Bell pair $|\Phi^+\rangle$ between some chosen destination stations A and B. We assume perfect classical communication among all stations but imperfect quantum operations and noisy channels, so that the local entangled pairs have limited fidelity; details of the error model are given in Sec. 5.2.2.

Let us assume, for the moment, that only bit-flip errors occur in the links of the network, with independent probability ε_b:

$$\rho = (1 - \varepsilon_b) |\Phi^+\rangle\langle\Phi^+| + \varepsilon_b |\Psi^+\rangle\langle\Psi^+|. \tag{5.1}$$

Phase errors are thus not considered yet, but they will be added in Sec. 5.2. In what follows, we show that one can use the links of the lattice to construct and propagate a large GHZ state through the lattice, so that a long-distance entangled pair of qubits can be generated (Sec. 5.1.1). The very special geometry of the network allows us to coherently check the parity of some pairs of qubits of the propagating GHZ state, leading to a pattern of *parity syndromes*. In the case of perfect connections, one gets only even-parity outcomes, but bit-flip errors introduce some random odd-parity results. However, we show in Sec. 5.1.2 that most bit-flip errors can be suppressed by a classical error correction based on the syndrome pattern.

5.1.1 Propagating a large GHZ state

The first step of the procedure consists of creating a local state $|\Phi^+\rangle$ at the corner station O, see Fig. 5.1. We then use the entangled pairs to teleport the two qubits of $|\Phi^+\rangle$ to the right and to the top until they reach C, in the following manner:

(i) Inner stations receive two qubits from their bottom and left neighbors, and they teleport them farther to the top and to the right, respectively.

(ii) Stations on a boundary do the same, but they either receive one qubit and "duplicate" it (by applying a CNOT with an additional qubit in the state $|0\rangle$ as target), or they receive

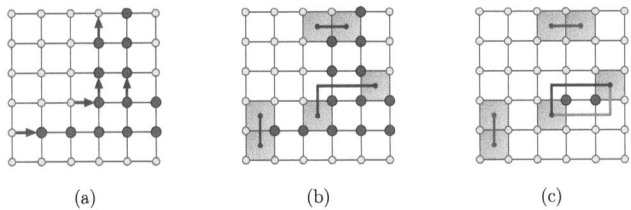

Figure 5.2: Bit-flip error correction. (a) Parity checks associated with five erroneous edges; nodes with odd parity are highlighted. (b) The corresponding plaquette representation. (c) In this example, the plaquettes are correctly paired but not all paths of errors are properly inferred. This creates a loop in the dual lattice, and stations lying inside this region apply the wrong bit-flip correction.

two qubits and "erase" one of them (by measuring it in the X basis and by communicating the outcome result). The station C erases the two qubits it receives.

(iii) The destination stations A and B save each an additional duplicated qubit, which are used to create the long-distance entangled state.

In a noiseless network, these operations generate a GHZ state that originates at O, propagates along the OC-diagonal, and expands and then shrinks in the perpendicular direction until it reaches C. During the expansion process, one indeed creates the state $|\text{GHZ}_{2t}\rangle$, where t denotes the time at which the teleportations are performed. Two observations are pertinent at this stage. First, one can include more qubits in the final state, creating a large GHZ state among the nodes of the lattice and not simply a Bell pair between only two stations. Second, more important, all teleportations can be executed at once. This sequence of operations is indeed only useful to visualize what is happening in the quantum network, but it does not play any role in the protocol. In fact, the choice of the local rotations that have to be applied on the final state, depending on the outcomes of the Bell measurements of the teleportations, can be postponed to the very end of the process; see also Fig. 5.4.

5.1.2 Network-based bit-flip error correction

A syndrome detection and a network-based error correction are now used to suppress most bit-flip errors that occur while teleporting the entangled state from O to C.

Parity checks and plaquette representation

The special geometry of the square lattice allows us to extract, without damaging the coherence of the final state, some information about the bit-flip errors. To this end, each station coherently computes the parity of the two qubits it receives, as shown in Fig. 5.4, and outputs $+1$ if the qubits have the same parity and -1 otherwise; we call this a *syndrome*. Since we are propagating a GHZ state, the presence of parity checks with the result -1 indicates the presence of bit-flip errors in the network. Remark that the stations at the bottom and left boundaries of the lattice cannot perform any parity check, since they receive one qubit only. As we will see in the next subsection, however, the fact that these syndromes are missing does not play any role in the efficiency of the error correction, as long as the stations A and B lie sufficiently far from the boundaries. For simplicity, let us therefore assume that the boundaries are error-free.

Given all parity checks of the lattice, the task is then to determine which are the edges responsible for the bit-flip errors. To this purpose, we attach to each plaquette of the network, that is, to each vertex of the dual lattice, a value that is the product of the four parity check outputs at its corners. For example, we show in Fig. 5.2a a possible realization of the protocol for a 6×6 lattice, in which only five teleportations flipped the qubits; the corresponding plaquette representation is depicted in Fig. 5.2b. One sees that single erroneous edges result in two neighboring plaquettes labeled by "-1", so that they can be identified unequivocally as long as they are *isolated*. However, having two (or more) such edges adjacent to the same plaquette leads to some ambiguity, see Fig. 5.2c. At this point, let us be more general and describe the optimal bit-flip correction. The error recovery runs as follows:

(i) Find all configurations of erroneous links that lead to the pattern of plaquette "-1" corresponding to the outcomes of the parity checks.

(ii) Randomly choose one of these configurations according to their probability of occurrence. Denoting by N_ε and $N_{\bar\varepsilon}$ the number of erroneous links and perfect ones, respectively, this probability is simply given by $\varepsilon_b^{N_\varepsilon}(1-\varepsilon_b)^{N_{\bar\varepsilon}}$.

(iii) Apply the bit-flip operator X on all qubits that are affected by the erroneous links of the chosen configuration.

Since this procedure is very similar, not to say identical, to the error recovery in surface codes (or more generally in topological quantum memories, see [DKLP02]), let us now compare both corrections and stress the differences that appear in the quantum correlations of the final state.

Comparison with the error recovery in surface codes

First, we consider the regime of small bit-flip error probabilities, so that the most probable configurations leading to a given pattern of plaquettes are the ones that contain the fewest errors. Because plaquettes appear in pairs (they are located at the extremities of paths of errors in the dual lattice, see Fig.5.2b), the most natural strategy is to find a coupling of the syndrome plaquettes that globally minimizes the distance between any two paired *defects* (plaquettes "-1"). In this way, one indeed finds a minimum set of erroneous edges that match the syndrome pattern.[2] The paired defects have then to be connected by a shortest path (in the dual lattice),

---
[2]Remark that this strategy is not optimal, since the degeneracy of the configurations with more errors may balance and even surpass their smaller probabilistic weight [SB09]. However, it yields some results that are close to optimality, especially for very small values of ε_b.

which is in general not unique. Since all such paths are equally probable, the choice of the correct one is ambiguous, and loops of wrongly inferred edges may be created, see Fig. 5.2c. In surface codes, if the error probability ε_b is too large, these loops proliferate and eventually give rise to a *homologically non-trivial loop* that stretches from one *rough edge* to another, suppressing any long-distance quantum correlation. In an infinite square lattice, this happens if ε_b is larger than the critical value

$$\varepsilon_b^* \approx 0.1094. \tag{5.2}$$

This threshold is determined via a mapping to the two-dimensional random-bond Ising model [HPP01] and corresponds to the optimum error correction. In our protocol, we do not deal with two opposite rough edges but rather four, since the whole network boundary lacks some information on the parity checks. This is the case because stations lying on the left and bottom edges of the lattice cannot compute any parity check (they receive only one qubit of the propagating GHZ state), while the links of the top and right edges create only one syndrome plaquette (and not two). However, paths of errors due to these boundary effects enter only superficially the lattice, such that the threshold for the error correction is not altered. In fact, the probability that such paths penetrate a distance d into the network decreases exponentially with d.

At this point, a basic difference between the two models has to be pointed out: in the surface code, homologically trivial loops do not affect the state used as a quantum memory, while non-contractible ones induce a logical error. In our quantum network, however, a contractible loop also affects the correlations of the final state if one of the destination stations lies in its inside, since, in this case, the wrong bit-flip error correction is applied to it. In that sense, the destination stations in our model can be viewed as punched holes in the planar code, which are used to encode logical qubits.

This observation removes part of the ambiguity concerning the choice of the path connecting two defects: it has to follow (as well as possible) a straight line. For the syndrome pattern shown in Fig. 5.2c, for instance, one easily verifies that the path "$\leftarrow\downarrow\leftarrow$" (starting from the rightmost plaquette) minimizes the average number $\bar{n}$ of sites that infer the wrong error correction:

$$\bar{n}(\leftarrow\downarrow\leftarrow) = \frac{2}{3}, \tag{5.3a}$$

while we have for the two other possible paths of minimum length:

$$\bar{n}(\downarrow\leftarrow\leftarrow) = \bar{n}(\leftarrow\leftarrow\downarrow) = 1. \tag{5.3b}$$

Fidelity of the resulting state

We have seen that no long-distance entanglement can be generated if $\varepsilon_b > \varepsilon_b^*$, so let us now study the fidelity F of the final state for smaller error rates. Since we assume bit-flip errors

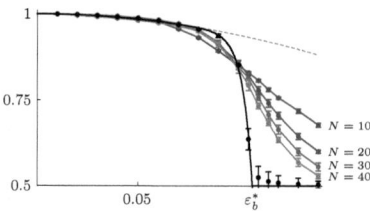

Figure 5.3: Monte Carlo simulations of the bit-flip error correction in a $N \times N$ square lattice. An unknown (classical) bit is sent through the noisy network, and a minimum set of erroneous edges leading to the corresponding parity check pattern is found using Edmonds' algorithm. In this plot, we show the probability $P_N(\varepsilon_b)$ that two random sites infer the same value of the bit if the bit-flip error probability is ε_b. The extrapolated function $P_\infty(\varepsilon_b)$ is plotted in bold line, while the dashed line represents its behavior for small error rates.

only, the quantum state shared by A and B is a mixture of the two Bell pairs $|\Phi^+\rangle$ and $|\Psi^+\rangle$:

$$\rho_{AB} = F\,|\Phi^+\rangle\langle\Phi^+| + (1-F)\,|\Psi^+\rangle\langle\Psi^+|. \tag{5.4}$$

This mixed state is entangled if $F > 1/2$ and can be distilled at a rate $E \equiv 1 - H_2(F)$, which is called distillable entanglement [BBP+96] and where the Shannon entropy H_2 is defined as

$$H_2(x) \equiv -x \log_2(x) - (1-x)\log_2(1-x). \tag{5.5}$$

Based on Edmonds' algorithms for finding a minimal weight perfect matching of the syndrome plaquettes [Edm65a, Edm65b], Monte-Carlo simulations were performed to compute the fidelity $F(\varepsilon_b)$, that is, the probability $P_\infty(\varepsilon_b)$ to apply the same bit-flip correction at arbitrarily distant stations A and B, see Fig. 5.3. For simplicity, and justified by the discussion on the rough edges in surface codes, we considered periodic boundary conditions for the network, so that syndrome plaquettes always appear in pairs. The presence of a threshold around 10.5% is confirmed, and the numerical results agree perfectly with the series expansion of F for $\varepsilon_b \ll 1$:

$$F(\varepsilon_b) = 1 - 6\,\varepsilon_b^2 + \mathcal{O}(\varepsilon_b^3). \tag{5.6}$$

Measurement errors

In practice, the parity check measurements are imperfect and yield a wrong result with probability ε_c, which lowers the threshold ε_b^*. The value of the new threshold can be estimated in a simple way by an entropy argument: each imperfect quantum channel introduces an entropy $H_2(\varepsilon_b)$ into the network, and each parity check extracts at most $1 - H_2(\varepsilon_c)$ bit of information. Therefore, an ordered phase can be maintained if

$$2H_2(\varepsilon_b) < 2H_2(\varepsilon_b^*) = 1 - H_2(\varepsilon_c). \tag{5.7}$$

Note that, for $\varepsilon_c = 0$, one finds $\varepsilon_b^* \approx 11\%$, which is very close to the real threshold, see Eq. (5.2). We now try to get rid of the measurement errors by repeating the parity checks $2r+1$ times and using the majority vote to infer the correct syndrome. If the parity check measurements do not perturb the qubits, repeating them can suppress the errors up to $\mathcal{O}(\varepsilon_c^{r+1})$. Even if additional errors are introduced into the system, three repeated measurements already help in correcting errors. In fact, a measurement error can be treated as an effective contribution to the error in the channel, see Sec. 5.2.2, which approximately becomes $\varepsilon_b' = \varepsilon_b + 3\varepsilon_c$, and an ordered phase is maintained whenever $\varepsilon_b' < \varepsilon_b^*$.

Other lattice geometries

Even if the square lattice is the best-adapted one for our purpose, ideas of network-based correction can easily be generalized to other regular lattices. The triangular lattice, for instance, has a slightly higher critical value: $\varepsilon_b^* \approx 17\%$. This threshold is found by solving the equation $3H_2(\varepsilon_b^*) = 2$ (similar entropy argument as for the square lattice), or by considering the corresponding random-bond Ising model on the triangular lattice [Ohz07].

5.2 A fault-tolerant protocol via encoding

We now propose a way of suppressing both bit-flip and phase errors that are present in the quantum network. Moreover, we show that all quantum operations can be performed fault tolerantly, that is, errors occurring on the qubits may be considered as independent.

Each physical qubit considered so far is replaced by an encoded block of qubits, and we also implement all quantum operations at the encoded level. Phase errors are suppressed by the code redundancy, and bit-flip errors are corrected exactly as explained in the previous section. The subspace for the redundant code of $n = 2t+1$ physical qubits is spanned by the two logical GHZ states

$$|\tilde{0}\rangle \equiv \frac{|+\rangle^{\otimes n} + |-\rangle^{\otimes n}}{\sqrt{2}} \quad \text{and} \quad |\tilde{1}\rangle \equiv \frac{|+\rangle^{\otimes n} - |-\rangle^{\otimes n}}{\sqrt{2}}. \quad (5.8)$$

This code can correct, by majority vote, up to t phase errors in a block of n qubits. However, it cannot correct any bit-flip errors, which we denote by $(t_p, t_b) = (t, 0)$. This is a CSS code with stabilizers generated by $\{X_1X_2, X_2X_3, \ldots, X_{n-1}X_n\}$, and all the nice properties of such codes can be used, as transversal CNOT gates or efficient measurements [Got98]. Furthermore, all the quantum operations discussed in the previous section can still be applied, with some minor changes, however:

(i) Physical qubits are replaced by encoded qubits. In particular, we use the encoded Bell pair $|\tilde{0}\tilde{0}\rangle + |\tilde{1}\tilde{1}\rangle$ as elementary links.

(ii) The CNOT gate is implemented by a transversal CNOT gate between two encoded qubits.

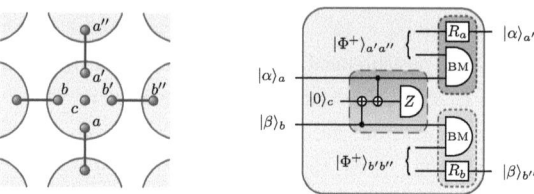

Figure 5.4: Local resources and quantum operations at each station (here, without encoding). At least five qubits are needed: one for the parity check measurement (dashed box), and four for the teleportations (dotted boxes). The choice of the rotations R, depending on the outcomes of the Bell measurements, can be tracked classically and thus no classical communication is necessary between the stations during the process. The interpretation of the parity checks can be determined in the end of the protocol.

(iii) Encoded Pauli operators $\tilde{X}$ and $\tilde{Z}$ are inferred by measuring all X and Z operators on the qubits of the encoding block.

(iv) A classical error correction is performed to suppress up to t phase errors.

It is important for the encoding process to fulfill the requirement of *fault tolerance*: the probability to get errors on j physical qubits should be of the order of ε_p^j for all $j \leq t$. Efficient procedures to fault-tolerantly prepare GHZ states are available, see [Kni05] and Sec. IX in [JTSL07], and therefore we can treat errors on physical qubits as being independent.

5.2.1 Required physical and temporal resources

As illustrated in Fig. 5.4 in the case of the trivial encoding $n = 1$, each station requires $4n$ qubits for the connections with its neighbors and n qubits for the parity check measurements, which can also be used to store the final Bell state. We therefore need approximately $5n$ qubits at each station, and the goal of this section is to relate n, as well as the maximum tolerable error rates ε_b^* and ε_p^*, to the size N of the lattice.

Tolerable error rates for a lattice of large but finite size

In order to estimate the fidelity of the long-distance entangled state that is created using our protocol, one has to quantify the amount of errors occurring at three different levels: let $\varepsilon \ll 1$ be the error probability at the physical level, $\tilde{\varepsilon}$ at the logical level, and ϵ at the network level. In Sec. 5.2.2, we find the approximation $\varepsilon \equiv \varepsilon_b \approx \varepsilon_p \lesssim 8.5\,\beta$, where β is the maximum error probability associated with the local two-qubit quantum gates, the measurements, and the quantum memory. At the encoded level, we have to distinguish the two types of errors, since the encoding preferentially suppresses the phase errors, while it moderately increases the bit-flip

N	10^1	10^2	10^3	10^4	10^5
$\varepsilon^*(E = 0.75)$	1.38	0.96	0.75	0.62	0.54
$\varepsilon^*(E = 0.50)$	1.84	1.27	0.97	0.79	0.67
$\varepsilon^*(E = 0.25)$	2.02	1.41	1.08	0.88	0.74

Table 5.1: Estimation of the resources that are required to create a long-distance pair of distillable entanglement E in a $N \times N$ square lattice. The size of the encoding is $n = 2t + 1$, and ε^* represents the elementary error probability (in percent) that can be tolerated. The total number of qubits at each station is $5n \approx 15 + 10 \log_{10}(N)$.

error rates:

$$\tilde{\varepsilon}_p = \sum_{j=t+1}^{2t+1} \binom{2t+1}{j} \varepsilon^j \approx \binom{2t+1}{t+1} \varepsilon^{t+1}, \quad (5.9a)$$

$$\tilde{\varepsilon}_b = \sum_{j=0}^{t} \binom{2t+1}{2j+1} \varepsilon^{2j+1} \approx (2t+1)\,\varepsilon. \quad (5.9b)$$

The phase errors accumulate as the $2N(N-1)$ links of the network are consumed for the teleportations, and the bit-flip correction is successfully applied with a probability $P_N(\tilde{\varepsilon}_b)$, see Fig. 5.3:

$$\epsilon_p = 1 - (1 - \tilde{\varepsilon}_p)^{2N(N-1)} \approx 2N^2\,\tilde{\varepsilon}_p, \quad (5.10a)$$

$$\epsilon_b = 1 - P_N(\tilde{\varepsilon}_b) \approx 1 - P_\infty(\tilde{\varepsilon}_b), \quad (5.10b)$$

where the approximations hold for $N \gg 1$. This finally results in the long-distance entangled state

$$\rho_{AB} = (1 - \epsilon_b)(1 - \epsilon_p)\,|\Phi^+\rangle\langle\Phi^+| + \epsilon_b(1 - \epsilon_p)\,|\Psi^+\rangle\langle\Psi^+| + \epsilon_p(1 - \epsilon_b)\,|\Phi^-\rangle\langle\Phi^-|$$
$$+ \epsilon_b \epsilon_p\,|\Psi^-\rangle\langle\Psi^-|. \quad (5.11)$$

By fixing, for example, the distillable entanglement $E(\rho_{AB}) \equiv 1 - H_2(\epsilon_b) - H_2(\epsilon_p)$ of the final state, and under the conditions that t is an integer and that ϵ_b and ϵ_p are of the same order of magnitude, one can estimate the required resources t (number of qubits used for the encoding) and ε (tolerable error probability at the physical level) by solving Eqs. (5.9) and (5.10), see Tab. 5.1. The number $n = 2t + 1$ of elementary links between neighboring nodes scales only logarithmically with the size of the lattice, and even though the maximum tolerable error rate ε^* decreases with N, it stays on the order of one percent for any quantum network of realistic size. More precisely, setting $n \sim \log(N)$, one finds that $E(\rho_{AB})$ tends to a strictly positive value if $n\varepsilon \approx \tilde{\varepsilon}_b \lesssim 11\%$. In fact, phase errors at the level of the network are completely suppressed in this case. It follows that ε^* decreases only logarithmically with N.

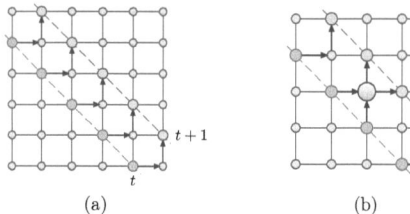

Figure 5.5: A one-dimensional quantum computer embedded in a lattice. (a) Two teleportations transport the quantum computer between times t and $t+1$. (b) Two-qubit gates can be applied on neighboring qubits.

Simultaneous measurements versus quantum memory

One key advantage of the proposed procedure is that all stations can operate *simultaneously*: the quantum operations at each station (including both Bell and parity-check measurements) can be made without knowing the measurement outcomes from the other stations. The interpretation of these outcomes, namely the choice of the local Pauli frames, is done by using one-way classical communication. Since the goal is to produce a Bell pair between the stations A and B, only the two concerned qubits have to be kept in quantum memory while waiting for the measurement outcomes to be collected and analyzed. At the same time, another round of quantum communication in the network can already start.

A more favorable application of our protocol is the distribution of quantum keys. To that purpose, all what is needed is the statistical correlations associated with the Bell pair, rather than the pair itself. This observation allows us to avoid any problem of decoherence of the qubits at stations A and B, since they can be measured even before the reception of all measurement outcomes. In fact, the two qubits are measured in one out of two complementary bases (for example X and Z), which is randomly chosen for each destination station, and the outcomes are secretly stored while the choice of the basis is publically announced. Once all measurement outcomes are received, A and B can determine whether they chose the same basis or not. This is the case with probability one-half, and thus a raw key for cryptography is available. Even if the correlation is not perfect due to various imperfections, we can apply the procedure of privacy amplification before we obtain the final highly correlated secret key (see p. 186 in [GRTZ02]).

Universal computation on a line

A more general question is whether entanglement distribution in a two-dimensional lattice, with a fixed local dimension and containing any kind of errors, is possible at all or not. Here, we show that this question is related to the existence of fault-tolerant quantum computation in a one-dimensional setting (two parallel lines of qubits) restricted to nearest-neighbor gates only.

Let us take one diagonal line of the lattice as a one-dimensional quantum computer at time $t = 0$, see Fig. 5.5a. We can translate this quantum computer in the upper-right direction by teleporting all its qubits to the right and then to the top. The quantum computer is now supposed to be at time $t = 1$, and the errors that occurred during the teleportations are seen as memory errors from time $t = 0$ to $t = 1$. The line can be further transported to reach any time t. Any two-qubit gate between neighboring qubits a and b can be implemented by slightly changing the path of the teleportations, as depicted in Fig. 5.5b: one of the qubit is teleported as usual, while the other is first teleported to the top and then to the right. In this way, the two qubits meet at the center station, at which the gate is applied. What we get is a nearest-neighbor one-dimensional quantum computation scheme with a simple error model, in which bit-flip and phase errors occur randomly with probabilities ε_b and ε_p, respectively.

In conclusion, the *spacelike* challenge of teleporting a qubit in a two-dimensional lattice is therefore replaced by the *timelike* task of preserving a qubit in a one-dimensional quantum computer. Since there exist such fault-tolerant quantum computation schemes using two qubits per site [SFH08], quantum information can be transported over arbitrary distances if one replaces all single-qubit links in our lattice by two-qubit links.

5.2.2 Towards a realistic scenario

In this section, we briefly study the different aspects of a possible implementation of our protocol. First, we explain how physical errors at the level of the qubits are related to the error rates ε_b and ε_p of our model. Then, we discuss the case of remote entanglement generated by photons, where their loss in the optical fibers has to be explicitly taken into account.

Error model

Let us describe a general error model independent of the physical realization of the system, in which the major errors are:

(i) The infidelity $1 - F_0$ of the elementary entangled pairs ρ, with $F_0 = \langle \Phi^+ | \rho | \Phi^+ \rangle$.

(ii) A local two-qubit gate error probability β and a local measurement error probability δ.

(iii) A memory error $\mu \approx \gamma T_0$ for a storage time T_0, assuming T_0 to be the time scale for generating encoded Bell pairs between neighboring stations.

We use the depolarizing channel for describing an error on a two-qubit gate O_{12}^{ideal} [BDCZ98]:

$$\rho \mapsto O_{12}[\rho] = (1 - \beta) O_{12}^{\text{ideal}}[\rho] + \frac{\beta}{4} \mathbb{1}_{12} \otimes \text{tr}_{12}[\rho], \tag{5.12}$$

while the following operators characterize an imperfect measurement on a single qubit:

$$P_0^\delta = (1 - \delta) |0\rangle\langle 0| + \delta |1\rangle\langle 1|,$$
$$P_1^\delta = (1 - \delta) |1\rangle\langle 1| + \delta |0\rangle\langle 0|. \tag{5.13}$$

Equivalently, we can model an imperfect measurement on one qubit by applying an effective depolarizing channel

$$O_1^{2\delta}[\rho] = (1 - 2\delta)\rho + \delta \mathbb{1}_1 \otimes \operatorname{tr}_1[\rho] \tag{5.14}$$

followed by a perfect measurement. Even though the error probability for this channel is 2δ, the probability to get a wrong measurement outcome is only δ. Finally, the result of an imperfect memory is modeled by a similar depolarizing channel with error probability μ. If the initial fidelity F_0 is not very close to unity, we may use the idea of (nested) entanglement pumping [DBCZ99] to efficiently pump the Bell pairs to a higher fidelity. This is only limited by the imperfections of the local operations and by the decoherence of the qubits and can lead to a fidelity $F_0' \approx 1 - \frac{5}{4}\beta - \frac{3}{4}\mu$, see [JTSL07].

With the condition that all operations are performed fault tolerantly, we can estimate the total error probability which is accumulated on an individual physical qubit during the creation of the long-distance entangled pair. First, the probabilities for bit-flip and phase errors[3] associated with the entanglement purification, the local encoding, the CNOT gate and the quantum teleportation is approximately given by $4\beta + 2\delta + \mu/2$. Then, m repeated parity check measurements may introduce some bit-flip and phase errors with probability $m\beta/2$. For $m = 3$ rounds, the effective measurement error probability is about $\beta^2/2 + 3(\beta + \delta)^2$. Since *one* measurement error is equivalent to *two* bit-flip errors in two connected edges, the measurement error of order $(\beta + \delta)^2$ can be conservatively counted as a bit-flip error of order $\beta + \delta$. Finally, if we assume $\beta \approx \delta \approx \mu$, we find that the accumulated probabilities for the bit-flip and phase errors are $\varepsilon_b \approx \varepsilon_p \lesssim 8.5\beta$.

Implementation with photons

At the moment, the most important scenario to implement our ideas is the one in which photons carry the quantum information. Just to give an example of how our scheme applies in that case, let us consider the setup of [CCGFZ99, MMO$^+$07], where entanglement is generated between two atoms by sending photons. The photons travel through optical fibers and interfere at a 50:50 beam-splitter, with the outputs detected by single-photon detectors. Post-selection on the detection events then ensures that remote entanglement is generated. However, since this is a probabilistic process, the time required to create all the links scales logarithmically with the size of the lattice,[4] so that efficient quantum memories are needed. This last requirement can be relaxed in two-dimensional quantum networks, nevertheless. In fact, one can fix a constant time (independent of N) for the preparation of the links and then create the separable state $|+\rangle\langle+|^{\otimes 2} + |-\rangle\langle-|^{\otimes 2} = |\Phi^+\rangle\langle\Phi^+| + |\Psi^+\rangle\langle\Psi^+|$ wherever the entanglement generation failed.

---

[3] We conservatively count errors associated with the Pauli operator Y as both bit-flip and phase errors.

[4] Suppose that each attempt to prepare a link fails with probability q. After r repeated attempts, this probability is reduced to q^r, and the L links of the network are successfully created with probability $(1 - q^r)^L$, which is exponentially close to 1 for $r > \ln(L)/\ln(1/q)$.

Then, long-distance entanglement is successfully generated with our protocol by "forgetting" the location of the faulty links, which results in a slightly higher bit-flip error probability only. Alternatively, one can implement an error correction for surface codes suffering loss and get better error thresholds, see [SBD09].

CHAPTER 6

Fidelity threshold for long-distance entanglement in cubic networks

In this chapter, we investigate the problem of generating entanglement over arbitrarily long distances in noisy quantum networks if the amount of physical resources is fixed (*i.e.*, it does not increase with the distance). We focus on three-dimensional regular lattices, where edges are full-rank mixed states of two qubits and where quantum operations can be applied perfectly. In contrast to the protocols designed for two-dimensional systems, we prove that entanglement can be established between two infinitely distant qubits if the fidelity of the connections is larger than a critical value F_c. Therefore, we show that a constant overhead of local resources is sufficient to achieve long-distance quantum communication.

The protocol starts by creating a thermal cluster state from the noisy links of the quantum network (Sec. 6.1). Then, in Sec. 6.2, we show that useful quantum correlations between two distant nodes can be extracted from the cluster state using LOCC only. To that end, all but the two distant qubits are measured in a basis that depends on their position in the lattice (Sec. 6.2.1). The measurement outcomes lead to a pattern of *error syndromes*, and a classical error correction is applied on the long-distance entangled pair, restoring the quantum correlations if the error rate is not too high, see Sec. 6.2.2. We provide an analytical upper bound on the maximum tolerable error rate, and in Sec. 6.2.3 we present a much more accurate estimate ($\approx 2.27\%$) based on Monte Carlo simulations.

6.1 Quantum networks and cluster states

In the previous chapter, we have shown how a large GHZ state can be created and propagated in a noisy square lattice. This state is robust against bit-flip errors if their rate is not too high, but it is very fragile against phase errors. In fact, any phase error destroys the coherence of a GHZ state. Therefore, an encoding of the qubits is required, which leads to a logarithmic scaling of the physical resources per node. We are now looking for a fidelity threshold in a lattice with single connections, that is, we want to create a multipartite state that has the ability to correct both bit-flip and phase errors. Three-dimensional cluster states thus arise as a natural choice. In fact, they are known to possess an intrinsic capability of error correction, so that long-range entanglement between two faces of an infinite noisy cubic cluster state is possible [RBH05]. The protocol described in this chapter is based on this very construction with two radical differences, however. First, the settings are distinct, and second, we allow only local quantum operations *everywhere* in the lattice.

Let us now describe the quantum network that is studied in this chapter. We consider a cubic

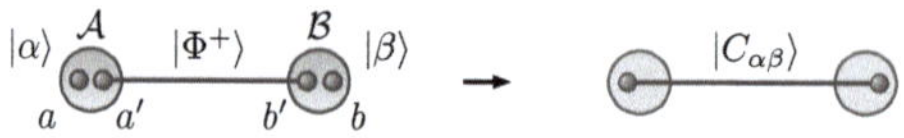

Figure 6.1: Non-local control phase on two qubits a and b using a Bell pair $|\Phi^+\rangle_{a'b'}$.

lattice that consists of N^3 sites and where neighboring nodes share one quantum connection subject to independent bit-flip and phase errors. For simplicity, we choose the error rates to be equal, *i.e.*, the bonds are described by the mixed states given in Eq. (4.13) with $\varepsilon = \varepsilon_b = \varepsilon_p$:

$$\rho = (1-\varepsilon)^2 |\Phi^+\rangle\langle\Phi^+| + \varepsilon(1-\varepsilon)\Big(|\Psi^+\rangle\langle\Psi^+| + |\Phi^-\rangle\langle\Phi^-|\Big) + \varepsilon^2 |\Psi^-\rangle\langle\Psi^-|.$$

In this setting, the nodes possess six qubits each (except the ones lying on the sides of the cube), on which arbitrary local quantum operations can be applied perfectly. We also assume that all classical processes (communication and computation) take much less time than any quantum operation. This last requirement is not crucial for the generation of long-distance entanglement, but it guarantees that the protocol runs in a time that does not scale with the network size.

Physical implementations of three-dimensional lattices have been proposed in the context of quantum information processing and distributed quantum computation [BCJD99, IM09]. For practical reasons, however, it may be advantageous to realize the proposed construction in two dimensions, using a "slice-by-slice" generation similar to the techniques developed in [RH07]. In that case, however, note that the time required to run the protocol scales linearly with N.

6.1.1 A mapping to noisy cluster states

A cluster state, which is an instance of graph states, can be constructed by inserting a qubit in the state $|+\rangle$ at each vertex of the graph and by applying a control phase between all neighboring pairs [HDR+06]. In our setting, we cannot perform these control phases since they are non-local quantum operations, but we can add an ancillary qubit and perform joint measurements at each node such that the result is a cluster state. This method was described in [VC04] in the case of perfect links, which can be interpreted as the virtual components of a large valence-bond state, and was then generalized to imperfect connections in [RBH05]. Nevertheless, let us describe here an explicit (and slightly different) construction, mainly for completeness sake but also for relating precisely the error rate in the quantum networks with that in the noisy cluster state.

At each node, we add a qubit $|+\rangle$ and use the noisy links ρ to indirectly perform the control phases. Let us first describe how this is achieved if all connections are perfect, *i.e.*, if they are in the state $|\Phi^+\rangle$. We consider two neighboring nodes A and B of the lattice, with two qubits a and b in the states $|\alpha\rangle = \sqrt{\alpha_0}|0\rangle + \sqrt{\alpha_1}|1\rangle$ and $|\beta\rangle = \sqrt{\beta_0}|0\rangle + \sqrt{\beta_1}|1\rangle$, and a connection $|\Phi^+\rangle$ between two qubits a' and b', see Fig. 6.1. We start by applying, on the qubits of A, the measurement operators

$$\begin{aligned}\mathcal{A}_0 &= |0\rangle_a\langle 00|_{aa'} + |1\rangle_a\langle 11|_{aa'},\\ \mathcal{A}_1 &= |0\rangle_a\langle 01|_{aa'} + |1\rangle_a\langle 10|_{aa'},\end{aligned} \qquad (6.1)$$

with $\sum_{i=0}^{1}\mathcal{A}_i^\dagger\mathcal{A}_i = \mathbb{1}_4$, which are followed by a bit-flip X on b' if the outcome is $\mathcal{A}_1$. The resulting state on a and b' reads $\sqrt{\alpha_0}\,|00\rangle + \sqrt{\alpha_1}\,|11\rangle$. We then apply a second measurement

$$\begin{aligned}\mathcal{B}_0 &= |0\rangle_b\langle +0|_{b'b} + |1\rangle_b\langle -1|_{b'b},\\ \mathcal{B}_1 &= |0\rangle_b\langle -0|_{b'b} + |1\rangle_b\langle +1|_{b'b},\end{aligned} \qquad (6.2)$$

followed by the matrix Z on the qubit a if we get $\mathcal{B}_1$ as outcome. Finally, the qubits a and b are left in the (entangled) state

$$|C_{\alpha\beta}\rangle \equiv \sqrt{\alpha_0\beta_0}\,|00\rangle + \sqrt{\alpha_0\beta_1}\,|01\rangle + \sqrt{\alpha_1\beta_0}\,|10\rangle - \sqrt{\alpha_1\beta_1}\,|11\rangle,$$

which is the result of a control phase between $|\alpha\rangle$ and $|\beta\rangle$. Clearly, if $|\alpha\rangle = |\beta\rangle = |+\rangle$, the state $|C_{\alpha\beta}\rangle$ is the cluster state on two qubits. Now, let us determine which errors occur if one blindly performs the very same operations but using another Bell state. It is straightforward to verify that one gets:

$$\begin{aligned}|\Psi^+\rangle &\to \mathbb{1}_2 \otimes Z \; |C_{\alpha\beta}\rangle,\\ |\Phi^-\rangle &\to Z \otimes \mathbb{1}_2 \; |C_{\alpha\beta}\rangle,\\ |\Psi^-\rangle &\to Z \otimes Z \; |C_{\alpha\beta}\rangle.\end{aligned} \qquad (6.3)$$

Since the matrices Z commute with the control phases, it follows that errors do not propagate while constructing a (noisy) cluster state ρ_{CS} from the quantum network. Moreover, because bit-flip and phase errors occur with independent probability ε in the links of the network, Z errors in the cluster state appear independently too. Since a node of the cubic lattice has degree six at most, and because two Z errors cancel each other, the vertices of ρ_{CS} suffer a Z error with probability smaller than or equal to

$$p = \sum_{i=0}^{2}\binom{6}{2i+1}\varepsilon^{2i+1}(1-\varepsilon)^{5-2i}. \qquad (6.4)$$

This expression reduces to $p \approx 6\varepsilon$ in the regime of small error rates. Therefore, we are exactly in the setting of [RBH05], where thermal fluctuations in the cluster state induce independent local Z errors with rate p. Remark that each node is now in possession of exactly one qubit.

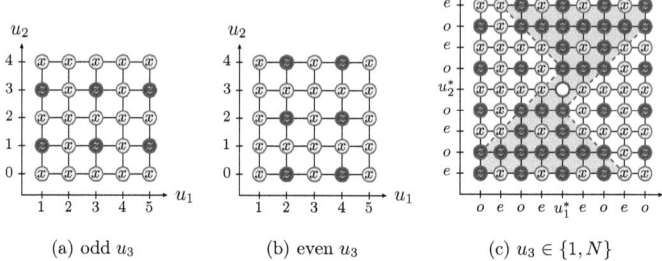

Figure 6.2: Mesurement pattern for the cluster state. a) Bases in which qubits in the bulk of a cluster state are measured; in this example $N = 5$. b) Slightly different measurement pattern for the faces $\mathcal{L}$ and $\mathcal{R}$: the central qubit is kept intact and all qubits that lie in the shaded area are measured in the Z basis, except the ones with coordinates $(e, e, 1)$ or (e, e, N), which are measured in the X basis.

6.2 Long-distance entanglement generation

In this section, we mainly follow the construction and the notation proposed in [RBH05], in particular the measurement of the qubits of ρ_{CS} according to a specific pattern of local bases. The outcomes of the measurements are random, but the choice of the bases establishes some parity constraints on them. Any violation of these constraints indicates an error, and a classical processing of all collected syndromes allows one to reliably identify the typical errors. This correction works perfectly for small error rates, but it breaks down at $p_c \approx 3.3\%$ [OAIM04]. The difference between the present method and that given in [RBH05] is that we do not allow any non-local quantum operation. This obliges us to design a more elaborated error correction, leading to a different type of long-distance entanglement. In fact, we are not going to create a pure and perfect Bell pair of logical qubits, but rather a mixture of two entangled physical qubits.

6.2.1 Measurement pattern and quantum correlations

Let us define a finite three-dimensional cluster state on the cube

$$\mathcal{C} = \{u = (u_1, u_2, u_3) : 1 \leq u_1, u_2 + 1, u_3 \leq N\},$$

and select two distant nodes A and B centered in two opposite faces $\mathcal{L}$ and $\mathcal{R}$. Their coordinates are $(u_1^*, u_2^*, 1)$ and (u_1^*, u_2^*, N), with $u_1^* = u_2^* + 1 = (N+1)/2$. For a reason that will become clear soon, we consider lattices of size $N \equiv 1 \pmod 4$, so that u_1^* is odd and u_2^* even. Let us also introduce two disjoint sublattices T_o and T_e with double spacing, where o and e stand for odd and even, respectively. Their vertices are

$$\begin{aligned} V(T_o) &= \{u = (o,o,o)\} \subset \mathcal{C}, \\ V(T_e) &= \{u = (e,e,e)\} \subset \mathcal{C}, \end{aligned} \tag{6.5}$$

and their edges are given by the sets

$$\begin{aligned} E(T_o) &= \{u = (o,o,e), (o,e,o), (e,o,o)\} \subset \mathcal{C}, \\ E(T_e) &= \{u = (e,e,o), (e,o,e), (o,e,e)\} \subset \mathcal{C}. \end{aligned} \tag{6.6}$$

We also define the planes

$$\begin{aligned} T_X^{(u_2)} &= \{u = (o, u_2, o)\} \subset T_o, \\ T_Z^{(u_1)} &= \{u = (u_1, e, e)\} \subset T_e, \end{aligned} \tag{6.7}$$

and denote by T_X^* and T_Z^* the planes that contain A and B. These planes will be used to derive the Bell correlations of the future long-distance entangled state $|\psi\rangle_{AB}$ (we first consider that no error occurs and then extend the results to noisy cluster states). Qubits that belong to the vertices of T_o and T_e are measured in the Z basis, while all other qubits are measured in the X basis, see Fig. 6.2. There are, however, some exceptions in $\mathcal{L}$ and $\mathcal{R}$. First, the central qubit is not measured, since it will be part of the long-distance entangled state. Second, qubits with coordinates $u_1 = u_1^*$ are measured in the Z basis in order to create the right quantum correlations, as explained in the following paragraph. Finally, we measure in the Z basis all qubits whose first two coordinates are (e,o) or (o,e) and which lie in the shaded areas; these outcomes will be important for the error correction.

To compute the effect of the measurements on the quantum correlations between A and B, we use the fact that a perfect cluster state $|C\rangle$ obeys the eigenvalue equation $K_u |C\rangle = |C\rangle$ for all $u \in \mathcal{C}$, where K_u is the stabilizer

$$K_u \equiv X_u \prod_{v \in \mathcal{N}(u)} Z_v, \tag{6.8}$$

and where $\mathcal{N}(u)$ stands for the neighborhood of u. If we let the products of stabilizers $\prod_{u \in T_X^*} K_u$ and $\prod_{u \in T_Z^*} K_u$ act on the cluster state, we find that A and B are maximally entangled:

$$\begin{aligned} X_A X_B |\psi_{AB}\rangle &= \lambda_X |\psi_{AB}\rangle, \\ Z_A Z_B |\psi_{AB}\rangle &= \lambda_Z |\psi_{AB}\rangle, \end{aligned} \tag{6.9}$$

with $\lambda_X, \lambda_Z \in \{-1, +1\}$. The eigenvalues $\lambda_{X,Z}$ are calculated from the measurement outcomes

x and z:

$$\lambda_X = \prod_{u \in \Omega_X^{(z)}} z_u \prod_{u \in \Omega_X^{(x)}} x_u,$$
$$\lambda_Z = \prod_{u \in \Omega_Z^{(z)}} z_u \prod_{u \in \Omega_Z^{(x)}} x_u, \qquad (6.10)$$

where the sets of vertices $\Omega_{X,Z}^{(x,z)}$ are given by

$$\Omega_X^{(x)} = T_X^* \setminus \{A, B\}, \qquad \Omega_X^{(z)} = T_X^{(u_2^* \pm 1)},$$
$$\Omega_Z^{(x)} = T_Z^*, \qquad \Omega_Z^{(z)} = T_Z^{(u_1^* \pm 1)} \cup \{(u_1^*, e, 1), (u_1^*, e, N)\} \setminus \{A, B\}.$$

6.2.2 Error correction and fidelity of the final state

As already mentioned, the measurement outcomes are random but not independent. It is in fact possible to assign to most vertices $u_i \in T_i$, with $i = o$ or e, the parity syndrome

$$s(u_i) \equiv \prod_{v \in \mathcal{N}(u_i)} K_v = \prod_{v \in \mathcal{N}(u_i)} x_v \prod_{w \in \mathcal{N}_i(u_i)} z_w, \qquad (6.11)$$

where $\mathcal{N}_i(u_i)$ designates the neighborhood of u_i in T_i. Since this equation arises from a product of stabilizers, we have that $s(u_i) = 1$ if no error occurs on the qubits of $\mathcal{N}_i(u_i)$. The key point of the construction is that a Z error on any edge of T_i changes the sign of the two syndromes at its extremities. This is due to the fact that Z errors do not commute with X measurements, while outcomes z are not affected by them. The sublattices T_o and T_e are treated separately, but in a similar way.

Syndrome-based error correction

We refer the reader to [DKLP02, RBH05] for a detailed discussion of the error recovery and present here only the basics to understand our protocol; remark that this error correction is essentially the same as the one described in the previous chapter. In the case of perfect and complete syndrome information, one knows exactly where all paths of Z errors start and end in each sublattice: the path extremities are located at the syndromes $s = -1$. In the regime of small error rate p, it turns out that a good recovery strategy is to pair these syndromes such that the total length of all pairings is minimized. Then, one connects any two paired syndromes by a path of minimum length and artificially introduces Z "errors" along these paths. This creates loops of errors in the cluster state, which, however, do not cause any damage to the long-distance quantum correlations. In fact, these loops either do not intersect the planes T_X^* and T_Z^* or cross them twice and consequently do not modify the eigenvalues in Eq. (6.9).

Problems arise because some syndromes are not known. For instance, consider the edges that have only one extremity in $V(T_o)$ or $V(T_e)$: their coordinates are $(o, 0, o)$ and $(o, N-1, o)$ in T_o,

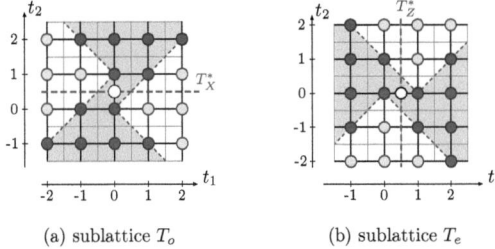

(a) sublattice T_o (b) sublattice T_e

Figure 6.3: Missing syndromes in the sublattices. a) In red (dark disks), missing syndromes at the vertices of T_o, in $\mathcal{L}$ and $\mathcal{R}$ (known syndromes are drawn in light gray). Rough faces lie on the top and the bottom of this lattice. A new coordinate system (t_1, t_2, t_3) is introduced for the vertices of the sublattice. b) Same considerations for the Z correlation: the missing syndromes create additional rough surfaces in T_e.

and $(1, e, e)$ and (N, e, e) in T_e, see Fig. 6.3. These are the rough faces described in [RBH05], and errors on these edges change the sign of only one syndrome (and not two) in the corresponding sublattice. An equivalent viewpoint is that both extremities of these edges indeed belong to T_o or T_e but that we do not have access to their outer syndrome. The consequence of this lack of information is that some paths of errors are not closed anymore, but rather originate from a missing syndrome and terminate at another. Typically, these open paths enter only superficially the lattice if the error rate p is small, but they start stretching from one side of the cube to another as soon as $p \gtrsim 3.3\%$. In the latter case, paths of errors cross an odd number of times the planes of correlations, which results in a complete loss of long-distance entanglement in the limit $N \to \infty$.

In contrast with [RBH05], and besides the rough faces present in any surface code, we also suffer a lack of syndrome information in $\mathcal{L}$ and $\mathcal{R}$. In fact, we cannot have a perfect and complete syndrome pattern for both T_o and T_e in these faces. For this to happen, one should be able to measure both x and z eigenvalues of the concerned qubits, which is impossible, or apply non-local quantum operations, which we do not allow. Actually, useful long-distance quantum correlations can still be created if one performs the measurements depicted in Fig. 6.2c: half outcomes are used to gain information on T_o, and symmetrically for T_e, see Fig. 6.3.

As an example of the effect of the unknown syndromes in $\mathcal{L}$, let us consider that an error occurred at the center station A, and that all other qubits did not suffer any error. Since we do not know the syndromes of T_o that lie directly below and above T_X^*, we are not able to restore the X correlation. This occurs with probability $p_X = p + \mathcal{O}(p^2)$. From this observation, one finds that the final state on A and B is a mixed state of the form

$$\rho_{AB} = F_X F_Z |\Phi^+\rangle\langle\Phi^+| + p_X F_Z |\Psi^+\rangle\langle\Psi^+| + F_X p_Z |\Phi^-\rangle\langle\Phi^-| + p_X p_Z |\Psi^-\rangle\langle\Psi^-|, \qquad (6.12)$$

with $F_X = 1 - p_X$ and $F_Z = 1 - p_Z$. This state is known to be distillable, and thus useful from a

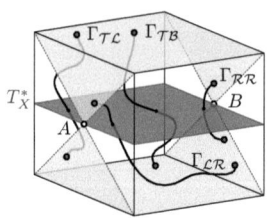

Figure 6.4: Some paths of errors that have a non-trivial effect on the long-distance entanglement: any path stretching from one shaded area to another and crossing the plane T_X^* an odd number of times degrades the X correlation between A and B. The shaded areas, which partially wrap the cube, are of two types: the top and bottom ones are the usual rough surfaces present in (three-dimensional) surface codes. The left and right shaded areas represent the unknown syndromes in T_o, see Fig. 6.3a. The situation for T_e is very similar (the picture is rotated by $90°$ around the AB axis), with the difference that all shaded areas are rough surfaces in that case.

quantum information perspective, whenever its fidelity $F_{AB} \equiv F_X F_Z$ is larger than one-half, see Sec. 4.2. This can be achieved when the error rate p is smaller than a threshold p_c. In the next paragraphs, we first prove a lower bound on this value: $p_c \gtrsim 1.17 \times 10^{-3}$. Then, in Sec. 6.2.3, we present numerical results showing that the real threshold is indeed much larger: $p_c \gtrsim 2.27\%$.

Effect of the missing syndromes on the fidelity F_{AB}

Let us first consider the correlation loss due to the missing syndromes in T_o. Paths of errors, which we generically denote by Γ, have a non-trivial effect on the X correlation if they cross the plane T_X^* an odd number of times, as depicted in Fig. 6.4. Moreover, the number l of errors which actually occur on a path Γ is at least $L/2$, where L denotes its length. This is the case because our error correction always leads to a minimum pairing of the syndromes $s = -1$. We now follow Chap. V in [DKLP02] to find an upper bound on the probability p_X of inferring the wrong quantum correlation:

$$p_X \leq 2 \sum_{\Gamma_{\mathcal{LL}}} \mathrm{prob}(\Gamma_{\mathcal{LL}}) + \sum_{\Gamma_{\mathcal{LR}}} \mathrm{prob}(\Gamma_{\mathcal{LR}}) + \sum_{\Gamma_{\mathcal{TB}}} \mathrm{prob}(\Gamma_{\mathcal{TB}}) + 4 \sum_{\Gamma_{\mathcal{TL}}} \mathrm{prob}(\Gamma_{\mathcal{TL}}), \qquad (6.13)$$

where $\mathcal{B}$ and $\mathcal{T}$ stand for the bottom and top faces, respectively. Note that we already took into account the symmetries of the problem in this expression. For convenience, let us introduce a new coordinate system (t_1, t_2, t_3) for the vertices of T_o, such that $-N_o \leq t_1 \leq N_o$, $-N_o < t_2 \leq N_o$, and $0 \leq t_3 \leq 2N_o$, with $N_o = (N-1)/4$, see also Fig. 6.3. In this coordinate system, paths of errors $\Gamma_{\mathcal{TL}}$ travel a distance $L \geq N_o$ and can start from N_o^2 distinct missing syndromes in T_o (lower triangle in $\mathcal{L}$). Because for each vertex there are, in a cubic lattice, at most $5^L/2$ self-avoiding walks pointing upward, we find the following bound on the last term of Eq. (6.13):

$$\sum_{\Gamma_{\mathcal{TL}}} \mathrm{prob}(\Gamma_{\mathcal{TL}}) \leq N_o^2 \sum_{L \geq N_o} \frac{5^L}{2} \sum_{l=\lceil L/2 \rceil}^{L} \binom{L}{l} p^l (1-p)^{L-l}, \qquad (6.14)$$

where $\lceil L/2 \rceil$ denotes the smallest integer not less than $L/2$. The sum over l, together with the binomial coefficients, counts all possible paths of errors that appear in a given walk. One can check that the bound tends to zero in the limit $N_o \to \infty$ if $10\sqrt{p(1-p)} < 1$, i.e., if $p \lesssim 1\%$. The same result holds for the paths $\Gamma_{\mathcal{L}\mathcal{R}}$ and $\Gamma_{\mathcal{T}\mathcal{B}}$; note that this value is about three times smaller than the real critical point for thermal cluster states in three dimensions. Similar considerations for the paths $\Gamma_{\mathcal{L}\mathcal{L}}$ finally yield, for $p < 1\%$:

$$p_X \leq 2 \sum_{\Gamma_{\mathcal{L}\mathcal{L}}} \text{prob}(\Gamma_{\mathcal{L}\mathcal{L}}) \leq 2 \sum_{t_2 \geq 1} 2 t_2 \sum_{L \geq t_2} \frac{5^L}{2} \sum_{l \geq \lceil L/2 \rceil} \binom{L}{l} p^l (1-p)^{L-l}. \quad (6.15)$$

This bound never tends to zero but still converges to a small value if p is small enough. Before computing a threshold for F_{AB}, however, we first have to consider the errors made in the other sublattice.

The loss of correlation in T_e is treated very similarly, since the measurement pattern is symmetric. Nonetheless, there is a subtle difference in this case: the missing syndromes do not lie on the vertices of the sublattice but rather on its outer edges $\mathcal{L}$ and $\mathcal{R}$. This creates additional parts of rough faces in T_e. It follows that the corresponding paths of errors have their first and last edges pointing in the t_3 direction, so that a slightly better bound for the error p_Z is derived:

$$p_Z \leq 2 \sum_{t_1 \geq 1} 2 t_1 \sum_{L \geq t_1 + 2} \frac{5^{L-2}}{2} \sum_{l \geq \lceil L/2 \rceil} \binom{L}{l} p^l (1-p)^{L-l}. \quad (6.16)$$

Combining the two bounds on p_X and p_Z, we find:

$$p \lesssim 1.17 \times 10^{-3} \Rightarrow F_{AB} > 1/2, \quad (6.17)$$

which corresponds, via Eq. (6.4), to an error rate $\varepsilon \lesssim 1.95 \times 10^{-4}$ in the initial connections. This value is quite small, mainly because our counting of paths of errors is very crude. Note that there are only few such paths of small length, and therefore this analytical bound could be increased by carefully computing its smallest orders in p. For instance, let us consider the series expansions of p_X at first order in p. One sees that only three edges of $\mathcal{L}$ may degrade the X correlation: these are the bonds in T_o that cross T_X^* and whose first coordinate t_1 belongs to $\{-1, 0, 1\}$, see Fig. 6.3a. At second order, one can check that the probability to infer the wrong X correlation due to the missing syndromes in $\mathcal{L}$ is $p_X^{\mathcal{L}} = 3p + 48p^2$. Therefore, by symmetry, the fidelity $F_X = 1 - p_X^{\mathcal{L}}(1 - p_X^{\mathcal{R}}) - p_X^{\mathcal{R}}(1 - p_X^{\mathcal{L}})$ reads

$$F_X = 1 - 6p - 78p^2 + \mathcal{O}(p^3). \quad (6.18)$$

It is easy to see that a single error in T_e cannot damage the Z correlation, and a careful counting of all configurations with two errors yields $p_Z^{\mathcal{L}} = 19p^2$. Consequently, we find

$$F_Z = 1 - 38p^2 + \mathcal{O}(p^3). \quad (6.19)$$

At this point, however, we prefer to turn to Monte Carlo simulations to find a much better estimate of the error threshold, while Eqs. (6.18) and (6.19) will be used to validate the algorithm.

6.2.3 Numerical estimation of the fidelity threshold

The situation is very similar for the two sublattices T_o and T_e, so let us consider the situation in which errors occur independently on each edge of a lattice with probability p, and where each vertex is assigned the value $+1$ if it is connected to an even number of erroneous edges and -1 otherwise. The latter vertices are referred as *syndromes*. As in Sec. 5.1.2, the error correction is based on the original algorithms by Edmonds [Edm65a, Edm65b] to find a minimum-weight perfect matching of the syndromes. Here, however, one has to take into account the effects of the boundaries (missing syndromes) on the error correction and cannot simply assume them to be error-free. In fact, the distance between the nodes A and B and their closest missing syndromes does not increase with N but rather stays constant. The usual way to deal with unknown syndromes is to include them directly into the perfect matching algorithm, as follows (see also Chap. 4 in [WFSH09]). Let $G = (V, E)$ be a weighted graph, where V is the set of syndromes and E is the set of edges connecting the vertices with a weight given by their distance in the lattice. First, we connect each vertex v of V to a new vertex v', which corresponds to its nearest missing syndrome. Second, we create zero-weight edges between all vertices v', so that when two vertices are paired, their respective missing syndromes can be paired at no cost. Then, an optimum perfect matching of the resulting weighted graph can be found efficiently, see [CR99] for a more recent algorithm than Edmonds'. While optimized algorithms [She90] based on the Delaunay triangulation can greatly reduce the cardinality of E for two-dimensional surface codes (with $|E| \propto |V|$ instead of $|E| \propto |V|^2$), which is very important for the efficiency of the perfect matching algorithms, it is not clear at the moment how this can be generalized to our three-dimensional setting. Presently, the number of edges in G scales as $\mathcal{O}(N^6)$, and therefore this method is not so "efficient" in practice.

In the following section, we propose a slightly different approach to the error correction with missing syndromes. First, given a syndrome pattern, we infer the value ± 1 of the unknown syndromes. Second, we proceed with the original matching algorithm. The advantage is that the weights between the syndromes satisfy now the triangular inequality (for which Edmonds' algorithm can be optimized), which is not the case with the general method.

Inferring the missing syndromes

We propose a very simple way of assigning the value $+1$ or -1 to the missing syndromes, so that a good approximation of the optimal configuration is found. To that end, it is helpful to consider a typical realization of a noisy cluster state in the regime of small error rates, as depicted in Fig. 6.5. The algorithm reads:

(i) Initialize all unknown syndromes to $+1$.

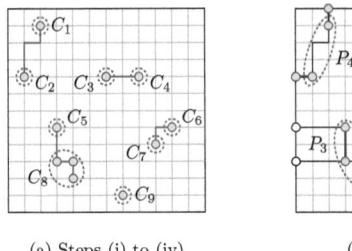

(a) Steps (i) to (iv) (b) Step (v)

Figure 6.5: An example of how unknown syndromes, which lie on the boundaries in this example, are assigned the value ± 1. a) All odd-size clusters C_i of syndromes -1 are paired by increasing distance. b) For each pair of clusters $P_k = \{C_i, C_j\}$, we check if it is favorable to add some new syndromes -1. Here, this is the case only for P_4 and P_5'.

(ii) Using nearest-neighbor site percolation, find all clusters C_i of syndromes -1. Keep the clusters of odd size only, and for each compute the minimum distance d_i to a closest unknown syndrome s_i.[1] We denote by n the number of such clusters. Note that we do not consider clusters of even size, since good pairings can be found for them, individually.

(iii) Calculate the distance d_{ij} between all pairs of clusters C_i and C_j, i.e., the length of the shortest path connecting C_i to C_j.

(iv) Find the indices a and b such that $d_{ab} = \min\{d_{ij}\}$, and create a pair $P_1 = \{C_a, C_b\}$. Remove C_a and C_b from the list of clusters and repeat the procedure until no cluster is left. In case of odd n, add one extra pair $\{C_i, C_i\}$ for the remaining cluster, with $d_{ii} = \infty$. This creates the list $\{P_1, \ldots, P_{n/2}\}$.

(v) For each $P_k = \{C_i, C_j\}$, check if $d_{ij} > d_i + d_j$. If this inequality holds, inverse the value of the corresponding missing syndromes: $s_i \leftarrow (-s_i)$ and $s_j \leftarrow (-s_j)$.

Error recovery

We now use Edmonds' algorithm to find an optimum pairing of the syndromes. The pairing is optimum in the sense that the total length of all pairs is minimized, that is, the fewest number of Z errors leading to the syndrome pattern is inferred, but note that certain matchings have a higher degeneracy than others, which may be taken into account, see the discussion in Sec. 5.1.2 or Ref. [SB09]. The error correction is successful if the parity of paths of errors crossing the plane of correlation is even. Simulations of the error corrections are performed for various lattice sizes (up to $\approx 15^3$ nodes) and for both X and Z correlations. The extrapolation to infinite lattices is done by fitting the data with an exponential function, see Fig. 6.6a. Final results are plotted

---

[1]Several such syndromes may exist. In that case, choose the one lying in the plane parallel to T_X^* or T_Z^* that contains C_i. This avoids unnecessary crossings of the plane of correlation.

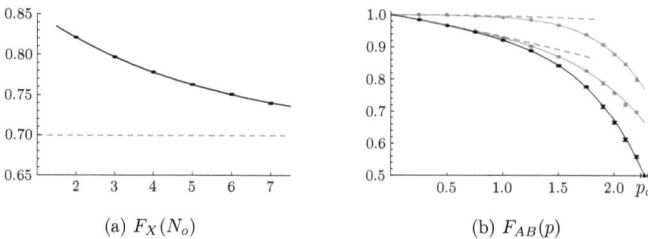

(a) $F_X(N_o)$ (b) $F_{AB}(p)$

Figure 6.6: Monte Carlo simulations. a) Fidelity F_X in the limit $N_o \to \infty$ for a fixed error rate p. We use the function $F_X^{(\infty)} + a\, e^{-b N_o}$ to fit the data computed from lattices consisting of $(2N_o + 1) \times (2N_o) \times (2N_o + 1)$ nodes. At least 10^5 simulations were run for each value of N_o. In this example, the error rate is $p = 2.2\%$ and the fit yields $F_X^{(\infty)} = 0.699 \pm 0.003$. b) Fidelity $F_{AB} = F_X F_Z$ as a function of the error rate p (in percent). The value $F_{AB} = 0.5$ is reached at $p_c = 2.27 \pm 0.03\,[\%]$. The upper curve represents the probability of success F_Z of the error correction in T_e, while the middle one is the function F_X. Corresponding series expansions for $p \ll 1$ are plotted with dashed lines.

in Fig. 6.6b: long-distance entanglement is achieved for error rates smaller than

$$p_c \approx 2.27\%, \tag{6.20}$$

or equivalently, $\varepsilon_c \approx 3.86 \times 10^{-3}$ for the original lattice. Let us now briefly comment on these thresholds:

(i) The values of the unknown syndromes are not optimally inferred in our algorithm, and therefore a higher value of p_c may be found. However, the series expansions plotted in Fig. 6.6b show that the proposed algorithm is optimal in the regime of very dilute errors, and it is clear that p_c cannot exceed the critical error rate of the usual three-dimensional thermal cluster states, so that in any case one finds $p_c \lesssim 3.3\%$.

(ii) One could get a higher threshold ε_c by computing directly F_{AB} as a function of ε. In fact, errors in the faces $\mathcal{L}$ and $\mathcal{R}$ do not appear with probability $p \approx 6\varepsilon$, but with probability $p \approx 5\varepsilon$ only.

(iii) Our measurement pattern puts the sublattices T_o and T_e on the same footing (Figs. 6.2 and 6.3), but it could be profitable to get more information on the unknown syndromes of T_o since the X correlation is more sensitive to errors.

(iv) As suggested in [RBH05], lattices of size $\log(N) \times \log(N) \times N$ may also be appropriate for generating long-distance entanglement. This result also holds in our setting, because additional errors only appear in the faces $\mathcal{L}$ and $\mathcal{R}$ and not in the bulk of the lattice.

Bibliography

[AB02] R. Albert and A.-L. Barabási. Statistical mechanics of complex networks. *Rev. Mod. Phys.*, 74(1):47–97, Jan 2002.

[ACL07] A. Acín, J. I. Cirac, and M. Lewenstein. Entanglement percolation in quantum networks. *Nature Phys.*, 3(4):256–259, Feb 2007.

[ADR82] A. Aspect, J. Dalibard, and G. Roger. Experimental test of Bell's inequalities using time-varying analyzers. *Phys. Rev. Lett.*, 49(25):1804–1807, Dec 1982.

[AGR81] A. Aspect, P. Grangier, and G. Roger. Experimental tests of realistic local theories via Bell's theorem. *Phys. Rev. Lett.*, 47(7):460–463, Aug 1981.

[BA99] A.-L. Barabási and R. Albert. Emergence of scaling in random networks. *Science*, 286:509–512, Oct 1999.

[BB84] C. H. Bennett and G. Brassard. Quantum cryptography: Public key distribution and coin tossing. In *Proceedings of IEEE International Conference on Computers, Systems and Signal Processing*, pages 175–179, New York, 1984.

[BBC+93] C. H. Bennett, G. Brassard, C. Crépeau, R. Jozsa, A. Peres, and W. K. Wootters. Teleporting an unknown quantum state via dual classical and Einstein-Podolsky-Rosen channels. *Phys. Rev. Lett.*, 70(13):1895–1899, Mar 1993.

[BBP+96] C. H. Bennett, G. Brassard, S. Popescu, B. Schumacher, J. A. Smolin, and W. K. Wootters. Purification of noisy entanglement and faithful teleportation via noisy channels. *Phys. Rev. Lett.*, 76(5):722–725, Jan 1996.

[BBPS96] C. H. Bennett, H. J. Bernstein, S. Popescu, and B. Schumacher. Concentrating partial entanglement by local operations. *Phys. Rev. A*, 53(4):2046–2052, Apr 1996.

[BCJD99] G. K. Brennen, C. M. Caves, P. S. Jessen, and I. H. Deutsch. Quantum logic gates in optical lattices. *Phys. Rev. Lett.*, 82(5):1060–1063, Feb 1999.

[BDCZ98] H. J. Briegel, W. Dür, J. I. Cirac, and P. Zoller. Quantum repeaters: The role of imperfect local operations in quantum communication. *Phys. Rev. Lett.*, 81(26):5932–5935, Dec 1998.

[BDJ09] S. Broadfoot, U. Dorner, and D. Jaksch. Entanglement percolation with bipartite mixed states. *EuroPhys. Lett.*, 88(5):50002, Dec 2009.

[BDSW96] C. H. Bennett, D. P. DiVincenzo, J. A. Smolin, and W. K. Wootters. Mixed-state entanglement and quantum error correction. *Phys. Rev. A*, 54(5):3824–3851, Nov 1996.

[Bel64] J. S. Bell. On the Einstein-Podolsky-Rosen paradox. *Physics*, 1(3), 1964.

[Bel04] J. S. Bell. *Speakable and Unspeakable in Quantum Mechanics*. Cambridge University Press, 2nd edition, 2004.

[Bol85] B. Bollobás. *Random graphs*. Academic Press, London, 1985.

[BR06] B. Bollobás and O. Riordan. *Percolation*. Cambridge University Press, 2006.

[BS03] S. Bornholdt and H. G. Schuster. *Handbook of graphs and networks: from the genome to the Internet*. Wiley-VCH, 2003.

[BVK98] S. Bose, V. Vedral, and P. L. Knight. Multiparticle generalization of entanglement swapping. *Phys. Rev. A*, 57(2):822–829, Feb 1998.

[BVK99] S. Bose, V. Vedral, and P. L. Knight. Purification via entanglement swapping and conserved entanglement. *Phys. Rev. A*, 60(1):194–197, Jul 1999.

[BW92] C. H. Bennett and S. J. Wiesner. Communication via one- and two-particle operators on Einstein-Podolsky-Rosen states. *Phys. Rev. Lett.*, 69(20):2881–2884, Nov 1992.

[CC09] M. Cuquet and J. Calsamiglia. Entanglement percolation in quantum complex networks. *Phys. Rev. Lett.*, 103(24):240503, Dec 2009.

[CCGFZ99] C. Cabrillo, J. I. Cirac, P. García-Fernández, and P. Zoller. Creation of entangled states of distant atoms by interference. *Phys. Rev. A*, 59(2):1025–1033, Feb 1999.

[CEHM99] J. I. Cirac, A. K. Ekert, S. F. Huelga, and C. Macchiavello. Distributed quantum computation over noisy channels. *Phys. Rev. A*, 59(6):4249–4254, Jun 1999.

[CHSH69] J. F. Clauser, M. A. Horne, A. Shimony, and R. A. Holt. Proposed experiment to test local hidden-variable theories. *Phys. Rev. Lett.*, 23(15):880–884, Oct 1969.

[CR99] W. Cook and A. Rohe. Computing minimum-weight perfect matchings. *INFORMS J. on Computing*, 11(2):138–148, Feb 1999.

[CTSL05] L. Childress, J. M. Taylor, A. S. Sørensen, and M. D. Lukin. Fault-tolerant quantum repeaters with minimal physical resources and implementations based on single-photon emitters. *Phys. Rev. A*, 72(5):052330, Nov 2005.

[CZC+07] Z.-B. Chen, B. Zhao, Y.-A. Chen, J. Schmiedmayer, and J.-W. Pan. Fault-tolerant quantum repeater with atomic ensembles and linear optics. *Phys. Rev. A*, 76(2):022329, Aug 2007.

[CZKM97] J. I. Cirac, P. Zoller, H. J. Kimble, and H. Mabuchi. Quantum state transfer and entanglement distribution among distant nodes in a quantum network. *Phys. Rev. Lett.*, 78(16):3221–3224, Apr 1997.

[DB07] W. Dür and H. J. Briegel. Entanglement purification and quantum error correction. *Rep. Prog. Phys.*, 70(8):1381–1424, Aug 2007.

[DBCZ99] W. Dür, H. J. Briegel, J. I. Cirac, and P. Zoller. Quantum repeaters based on entanglement purification. *Phys. Rev. A*, 59(1):169–181, Jan 1999.

[DEJ+96] D. Deutsch, A. Ekert, R. Jozsa, C. Macchiavello, S. Popescu, and A. Sanpera. Quantum privacy amplification and the security of quantum cryptography over noisy channels. *Phys. Rev. Lett.*, 77(13):2818–2821, Sep 1996.

[Deu85] D. Deutsch. Quantum theory, the Church-Turing principle and the universal quantum computer. *Proc. R. Soc. Lond. A*, 400(1818):97–117, Jul 1985.

[Die05] R. Diestel. *Graph theory*. Springer-Verlag Heidelberg, New York, 2005.

[Dij59] E. W. Dijkstra. A note on two problems with connections in graphs. *Numer. Math.*, 1:269–271, 1959.

[DiV95] D. P. DiVincenzo. Quantum computation. *Science*, 270(5234):255–261, Oct 1995.

[DKLP02] E. Dennis, A. Kitaev, A. Landahl, and J. Preskill. Topological quantum memory. *J. Math. Phys.*, 43:4452–4505, Sep 2002.

[DLCZ01] L.-M. Duan, M. D. Lukin, J. I. Cirac, and P. Zoller. Long-distance quantum communication with atomic ensembles and linear optics. *Nature*, 414:413–418, Nov 2001.

[dRMT+04] H. de Riedmatten, I. Marcikic, W. Tittel, H. Zbinden, D. Collins, and N. Gisin. Long distance quantum teleportation in a quantum relay configuration. *Phys. Rev. Lett.*, 92(4):047904, Jan 2004.

[Edm65a] J. Edmonds. Maximum matching and a polyhedron with 0,1-vertices. *J. Res. Nat. Bur. Stand.*, 69B:125–130, 1965.

[Edm65b] J. Edmonds. Paths, trees, and flowers. *Canad. J. Math.*, 17:449–467, 1965.

[Ein71] A. Einstein. *The Born-Einstein Letters; Correspondence between Albert Einstein and Max and Hedwig Born from 1916 to 1955*. Walker, New York, 1971.

[Eke91] A. K. Ekert. Quantum cryptography based on Bell's theorem. *Phys. Rev. Lett.*, 67(6):661–663, Aug 1991.

[EPR35] A. Einstein, B. Podolsky, and N. Rosen. Can quantum-mechanical description of physical reality be considered complete? *Phys. Rev.*, 47(10):777–780, May 1935.

[ER59] P. Erdős and A. Rényi. On random graphs. *Publ. Math. Debrecen*, 6:290–297, 1959.

[ER60] P. Erdős and A. Rényi. On the evolution of random graphs. *Publ. Math. Inst. Hung. Acad. Sci.*, 5:17–61, 1960.

[ER61] P. Erdős and A. Rényi. On the evolution of random graphs. ii. *Bull. Inst. Int. Stat.*, 38(4):343–347, 1961.

[FC72] S. J. Freedman and J. F. Clauser. Experimental test of local hidden-variable theories. *Phys. Rev. Lett.*, 28(14):938–941, Apr 1972.

[Fey82] R. P. Feynman. Simulating physics with computers. *Int. J. of Theor. Phys.*, 21(6):467–488, Jun 1982.

[FKG71] C. M. Fortuin, P. W. Kasteleyn, and J. Ginibre. Correlation inequalities on some partially ordered states. *Comm. Math. Phys.*, 22(2):89–103, 1971.

[FT76] E. S. Fry and R. C. Thompson. Experimental test of local hidden-variable theories. *Phys. Rev. Lett.*, 37(8):465–468, Aug 1976.

[FUH+09] A. Fedrizzi, R. Ursin, T. Herbst, M. Nespoli, R. Prevedel, T. Scheidl, F. Tiefenbacher, T. Jennewein, and A. Zeilinger. High-fidelity transmission of entanglement over a high-loss free-space channel. *Nature Phys.*, 5:389–392, May 2009.

[GFE09] D. Gross, S. T. Flammia, and J. Eisert. Most quantum states are too entangled to be useful as computational resources. *Phys. Rev. Lett.*, 102(19):190501, May 2009.

[Got98] D. Gottesman. Theory of fault-tolerant quantum computation. *Phys. Rev. A*, 57(1):127–137, Jan 1998.

[Gri99] G. Grimmett. *Percolation*. Springer-Verlag, Berlin, 1999.

[Gro96] L. Grover. A fast quantum mechanical algorithm for database search. In *Proceedings of the 28th Annual ACM Symposium on the Theory of Computing (STOC)*, pages 212–Ű219, May 1996.

[GRTZ02] N. Gisin, G. Ribordy, W. Tittel, and H. Zbinden. Quantum cryptography. *Rev. Mod. Phys.*, 74(1):145–195, Mar 2002.

Bibliography

[GS86] B. Grünbaum and G. C. Shephard. *Tilings and patterns*. W. H. Freeman & Co., New York, 1986.

[HDR+06] M. Hein, W. Dür, R. Raussendorf, M. Van den Nest, and H. J. Briegel. Entanglement in graph states and its applications. In *Quantum Computer, Algorithms and Chaos*, volume 162 of *International School of Physics Enrico Fermi*. IOS, Amsterdam, Oct 2006. edited by G. Casati, D. L. Shepelyansky, P. Zoller, and G. Benenti.

[HKBD07] L. Hartmann, B. Kraus, H. J. Briegel, and W. Dür. Role of memory errors in quantum repeaters. *Phys. Rev. A*, 75(3):032310, Mar 2007.

[HPP01] A. Honecker, M. Picco, and P. Pujol. Universality class of the Nishimori point in the 2d $\pm j$ random-bond Ising model. *Phys. Rev. Lett.*, 87(4):047201, Jul 2001.

[HS00] L. Hardy and D. D. Song. Entanglement-swapping chains for general pure states. *Phys. Rev. A*, 62(5):052315, Oct 2000.

[HSP10] K. Hammerer, A. S. Sørensen, and E. S. Polzik. Quantum interface between light and atomic ensembles. *Rev. Mod. Phys.*, 82(2):1041–1093, Apr 2010.

[HW97] S. Hill and W. K. Wootters. Entanglement of a pair of quantum bits. *Phys. Rev. Lett.*, 78(26):5022–5025, Jun 1997.

[IM09] R. Ionicioiu and W. J. Munro. Constructing 2D and 3D cluster states with photonic modules, 2009. eprint arXiv:quant-ph/0906.1727.

[Jan02] E. Jané. Purification of two-qubit mixed states. *Quant. Inf. Comput.*, 2(5):348–354, Aug 2002.

[JKBH08] L. Jahnke, J. W. Kantelhardt, R. Berkovits, and S. Havlin. Wave localization in complex networks with high clustering. *Phys. Rev. Lett.*, 101(17):175702, Oct 2008.

[JTL07] L. Jiang, J. M. Taylor, and M. D. Lukin. Fast and robust approach to long-distance quantum communication with atomic ensembles. *Phys. Rev. A*, 76(1):012301, Jul 2007.

[JTN+09] L. Jiang, J. M. Taylor, K. Nemoto, W. J. Munro, R. Van Meter, and M. D. Lukin. Quantum repeater with encoding. *Phys. Rev. A*, 79(3):032325, Mar 2009.

[JTSL07] L. Jiang, J. M. Taylor, A. S. Sørensen, and M. D. Lukin. Distributed quantum computation based on small quantum registers. *Phys. Rev. A*, 76(6):062323, Dec 2007.

[KC04] B. Kraus and J. I. Cirac. Discrete entanglement distribution with squeezed light. *Phys. Rev. Lett.*, 92(1):013602, Jan 2004.

[Ken98]   A. Kent. Entangled mixed states and local purification. *Phys. Rev. Lett.*, 81(14):2839–2841, Oct 1998.

[Kim08]   H. J. Kimble. The quantum Internet. *Nature*, 453:1023–1030, Jun 2008.

[KL96]   E. Knill and R. Laflamme. Concatenated quantum codes, Aug 1996. eprint arXiv:quant-ph/9608012.

[KLM01]   E. Knill, R. Laflamme, and G. J. Milburn. A scheme for efficient quantum computation with linear optics. *Nature*, 409:46–52, Jan 2001.

[Kni05]   E. Knill. Quantum computing with realistically noisy devices. *Nature*, 434:39–44, Mar 2005.

[Koc89]   M. Kochen. *The small world*. Ablex, Norwood, NJ, 1989.

[KRE07]   K. Kieling, T. Rudolph, and J. Eisert. Percolation, renormalization, and quantum computing with nondeterministic gates. *Phys. Rev. Lett.*, 99(13):130501, Sep 2007.

[LBAW08] B. P. Lanyon, M. Barbieri, M. P. Almeida, and A. G. White. Experimental quantum computing without entanglement. *Phys. Rev. Lett.*, 101(20):200501, Nov 2008.

[LMP98]   N. Linden, S. Massar, and S. Popescu. Purifying noisy entanglement requires collective measurements. *Phys. Rev. Lett.*, 81(15):3279–3282, Oct 1998.

[LWL09]   G. J. Lapeyre Jr., J. Wehr, and M. Lewenstein. Enhancement of entanglement percolation in quantum networks via lattice transformations. *Phys. Rev. A*, 79(4):042324, Apr 2009.

[LZ98]   C. D. Lorenz and R. M. Ziff. Precise determination of the bond percolation thresholds and finite-size scaling corrections for the sc, fcc, and bcc lattices. *Phys. Rev. E*, 57(1):230–236, Jan 1998.

[MG08]   J. Modławska and A. Grudka. Nonmaximally entangled states can be better for multiple linear optical teleportation. *Phys. Rev. Lett.*, 100(11):110503, Mar 2008.

[MHPC06] C. A. Muschik, K. Hammerer, E. S. Polzik, and J. I. Cirac. Efficient quantum memory and entanglement between light and an atomic ensemble using magnetic fields. *Phys. Rev. A*, 73(6):062329, Jun 2006.

[MMO$^+$07] D. L. Moehring, P. Maunz, S. Olmschenk, K. C. Younge, D. N. Matsukevich, L.-M. Duan, and C. Monroe. Entanglement of single-atom quantum bits at a distance. *Nature*, 449:68–71, Sep 2007.

[NBW06]   M. Newman, A.-L. Barabási, and D. J. Watts. *The structure and dynamics of networks*. Princeton Univ. Press, 2006.

[NC00]   M. A. Nielsen and I. L. Chuang. *Quantum Computation and Quantum Information.* Cambridge University Press, 2000.

[Nie99]   M. A. Nielsen. Conditions for a class of entanglement transformations. *Phys. Rev. Lett.*, 83(2):436–439, Jul 1999.

[NMW08]   R. A. Neher, K. Mecke, and H. Wagner. Topological estimation of percolation thresholds. *J. Stat. Mech.*, page P01011, Jan 2008.

[NV01]   M. A. Nielsen and G. Vidal. Majorization and the interconversion of bipartite systems. *Quant. Inf. Comput.*, 1(1):76–93, Jul 2001.

[OAIM04]   T. Ohno, G. Arakawa, I. Ichinose, and T. Matsui. Phase structure of the random-plaquette Z_2 gauge model: accuracy threshold for a toric quantum memory. *Nucl. Phys. B*, 697(3):462–480, Oct 2004.

[Ohz07]   M. Ohzeki. Multicritical points of Potts spin glasses on the triangular lattice. *J. Phys. Soc. Jpn*, 76(11):114003, Oct 2007.

[Par04]   R. Parviainen. *Connectivity Properties of Archimedean and Laves Lattices.* PhD thesis, Uppsala University, 2004.

[PCA+08]   S. Perseguers, J. I. Cirac, A. Acín, M. Lewenstein, and J. Wehr. Entanglement distribution in pure-state quantum networks. *Phys. Rev. A*, 77(2):022308, Feb 2008.

[PCL+10]   S. Perseguers, D. Cavalcanti, G. J. Lapeyre, M. Lewenstein, and A. Acín. Multipartite entanglement percolation. *Phys. Rev. A*, 81(3):032327, Mar 2010.

[Per10]   S. Perseguers. Fidelity threshold for long-range entanglement in quantum networks. *Phys. Rev. A*, 81(1):012310, Jan 2010.

[PGU+03]   J.-W. Pan, S. Gasparoni, R. Ursin, G. Weihs, and A. Zeilinger. Experimental entanglement purification of arbitrary unknown states. *Nature*, 423:417–422, May 2003.

[PJS+08]   S. Perseguers, L. Jiang, N. Schuch, F. Verstraete, M. D. Lukin, J. I. Cirac, and K. G. H. Vollbrecht. One-shot entanglement generation over long distances in noisy quantum networks. *Phys. Rev. A*, 78(6):062324, Dec 2008.

[PLAC10]   S. Perseguers, M. Lewenstein, A. Acín, and J. I. Cirac. Quantum random networks. *Nature Phys.*, doi:10.1038/nphys1665, May 2010.

[PPM08]   A. Poppe, M. Peev, and O. Maurhart. Outline of the SECOQC quantum-key-distribution network in Vienna. *Int. J. Quantum Inf.*, 6(2):209–218, Apr 2008.

[PT00]   A. Pönitz and P. Tittman. Improved upper bounds for self-avoiding walks in $\mathbb{Z}^d$. *Electronic J. Combinatorics*, 7(1):R21, Apr 2000.

[RBH05] R. Raussendorf, S. Bravyi, and J. Harrington. Long-range quantum entanglement in noisy cluster states. *Phys. Rev. A*, 71(6):062313, Jun 2005.

[RH07]   R. Raussendorf and J. Harrington. Fault-tolerant quantum computation with high threshold in two dimensions. *Phys. Rev. Lett.*, 98(19):190504, May 2007.

[SA91]   D. Stauffer and A. Aharony. *Introduction to Percolation Theory*. Taylor and Francis, London, 1991.

[SB09]   T. M. Stace and S. D. Barrett. Error correction and degeneracy in surface codes suffering loss, 2009. eprint arXiv:quant-ph/0912.1159.

[SBC09] A. K. Sen, K. K. Bardhan, and B.K. Chakrabarti (Eds.). *Quantum and Semiclassical Percolation and Breakdown in Disordered Solids*. Lecture Notes in Physics, Vol. 762. Springer, Berlin, 2009.

[SBD09] T. M. Stace, S. D. Barrett, and A. C. Doherty. Thresholds for topological codes in the presence of loss. *Phys. Rev. Lett.*, 102(20):200501, May 2009.

[SdA+07] C. Simon, H. de Riedmatten, M. Afzelius, N. Sangouard, H. Zbinden, and N. Gisin. Quantum repeaters with photon pair sources and multimode memories. *Phys. Rev. Lett.*, 98(19):190503, May 2007.

[SE63]   M. F. Sykes and J. W. Essam. Some exact critical percolation probabilities for bond and site problems in two dimensions. *Phys. Rev. Lett.*, 10(1):3–4, Jan 1963.

[SE64]   M. F. Sykes and J. W. Essam. Exact critical percolation probabilities for site and bond problems in two dimensions. *J. Math. Phys.*, 5(8):1117–1127, Aug 1964.

[SFH08] A. M. Stephens, A. G. Fowler, and L. C. L. Hollenberg. Universal fault tolerant quantum computation on bilinear nearest neighbor arrays. *Quantum Inf. Comput.*, 8(3&4):330–344, Mar 2008.

[She96] J. R. Shewchuk. Triangle: Engineering a 2D quality mesh generator and Delaunay triangulator, May 1996. From the First ACM Workshop on Applied Computational Geometry.

[Sho94] P. W. Shor. Algorithms for quantum computation: Discrete logarithms and factoring. In *Proceedings of the 35th Annual Symposium on Foundations of Computer Science*, pages 124–Ű134. IEEE Computer Society Press, Los Alamitos, CA, 1994.

[SSdRG09] N. Sangouard, C. Simon, H. de Riedmatten, and N. Gisin. Quantum repeaters based on atomic ensembles and linear optics, 2009. eprint arXiv:quant-ph/0906.2699.

[SSZ+08] N. Sangouard, C. Simon, B. Zhao, Y.-A. Chen, H. de Riedmatten, J.-W. Pan, and N. Gisin. Robust and efficient quantum repeaters with atomic ensembles and linear optics. *Phys. Rev. A*, 77(6):062301, Jun 2008.

[SZ99] P. N. Suding and R. M. Ziff. Site percolation thresholds for Archimedean lattices. *Phys. Rev. E*, 60(1):275–283, Jul 1999.

[VC04] F. Verstraete and J. I. Cirac. Valence-bond states for quantum computation. *Phys. Rev. A*, 70(6):060302(R), Dec 2004.

[Vid99] G. Vidal. Entanglement of pure states for a single copy. *Phys. Rev. Lett.*, 83(5):1046–1049, Aug 1999.

[VMDC04] F. Verstraete, M. A. Martín-Delgado, and J. I. Cirac. Diverging entanglement length in gapped quantum spin systems. *Phys. Rev. Lett.*, 92(8):087201, Feb 2004.

[VW01] K. G. H. Vollbrecht and R. F. Werner. Entanglement measures under symmetry. *Phys. Rev. A*, 64(6):062307, Nov 2001.

[Wat04] D. J. Watts. *Small worlds: the dynamics of networks between order and randomness*. Princeton Univ. Press, 2004.

[Wer89] R. F. Werner. Quantum states with Einstein-Podolsky-Rosen correlations admitting a hidden-variable model. *Phys. Rev. A*, 40(8):4277–4281, Oct 1989.

[WFSH09] D. S. Wang, A. G. Fowler, A. M. Stephens, and L. C. L. Hollenberg. Threshold error rates for the toric and surface codes, 2009. eprint arXiv:quant-ph/0905.0531.

[Wie83] S. Wiesner. Conjugate coding. *SIGACT News*, 15(1):78–88, 1983.

[Woo98] W. K. Wootters. Entanglement of formation of an arbitrary state of two qubits. *Phys. Rev. Lett.*, 80(10):2245–2248, Mar 1998.

[WS98] D. J. Watts and S. H. Strogatz. Collective dynamics of small-world networks. *Nature*, 393:440–442, Jun 1998.

[WZ82] W. K. Wootters and W. H. Zurek. A single quantum cannot be cloned. *Nature*, 299:802–803, Oct 1982.

[YCZ+08] Z.-S. Yuan, Y.-A. Chen, B. Zhao, S. Chen, J. Schmiedmayer, and J.-W. Pan. Experimental demonstration of a BDCZ quantum repeater node. *Nature*, 454:1098–1101, Aug 2008.

[ZZHE93] M. Żukowski, A. Zeilinger, M. A. Horne, and A. K. Ekert. "Event-ready-detectors" Bell experiment via entanglement swapping. *Phys. Rev. Lett.*, 71(26):4287–4290, Dec 1993.

Acknowledgment

First and foremost, I would like to thank my advisor J. Ignacio Cirac for the helpful advices and suggestions he has given to me over the course of my doctoral studies. I have profited greatly not only from his faculty to discern the importance and to reach the very essence of scientific questions, but also from his optimistic confidence in the possibility to answer them. His ability to speak about physics in a deep and rigorous but nevertheless extremely clear and enthusiastic manner is inestimable for a Ph.D. student. I will always be indebted to him for these immense scientific and human qualities.

I express my sincerest gratitude to Antonio Acín and Maciej Lewenstein for their warm hospitality during my three-month stay at the ICFO, and for the passionate discussions we have enjoyed in many other places. More than a fruitful collaboration on several chapters of this Thesis, working with them has been a constant source of inspiration, motivation, and cheerfulness.

I heartily thank the numerous people who contributed to broadening my knowledge on quantum information. I would like to thank especially Géza Giedke, who tirelessly responded to my questions during the first year of my doctoral studies, and Karl-Gerd Vollbrecht, who took over this job the following year. I have also profited from many sporadic but nonetheless very interesting discussions with Miguel Aguado, Daniel Cavalcanti, Alastair Kay, G. John Lapeyre, Maarten van den Nest, Roman Schmied, and Norbert Schuch.

In quality of coordinator of the Ph.D. program "QCCC", Thomas Schulte-Herbrüggen has devoted a considerable energy in animating seminars, organizing workshops, or proposing topics of wide interest. I thank him and my fellow students for the great time we have spent together.

Since my very first day at the MPQ, I have had the opportunity to share the office with Mikel Sanz, a lively and brilliant compeer. Besides the many creative discussions we have had, his patience in improving my spoken Spanish has been impressive, and priceless.

My regards extend to all my colleagues and friends, not only in Munich but also in Barcelona and in my home country. I do not even try to enumerate them; this could only results in a very incomplete and restrictive list. The persons closest to me know perfectly that their friendship has been a key ingredient for achieving this Thesis, anyway. I am sure that they will forgive me not to put their name on this very last page.

Finally, I cannot express enough my warmest thanks to my family, my sister, my brother, and my parents. For years, they have supported me in all possible ways, making my student life as enjoyable and comfortable as possible. Their genuine curiosity about my work and their continuous affection and encouragement have brightened many days of my living abroad. I hope I am able to return the favor, even just a little bit.

Die VDM Verlagsservicegesellschaft sucht für wissenschaftliche Verlage abgeschlossene und herausragende

Dissertationen, Habilitationen, Diplomarbeiten, Master Theses, Magisterarbeiten usw.

für die kostenlose Publikation als Fachbuch.

Sie verfügen über eine Arbeit, die hohen inhaltlichen und formalen Ansprüchen genügt, und haben Interesse an einer honorarvergüteten Publikation?

Dann senden Sie bitte erste Informationen über sich und Ihre Arbeit per Email an *info@vdm-vsg.de*.

Sie erhalten kurzfristig unser Feedback!

VDM Verlagsservicegesellschaft mbH
Dudweiler Landstr. 99
D - 66123 Saarbrücken

Telefon +49 681 3720 174
Fax   +49 681 3720 1749

www.vdm-vsg.de

Die VDM Verlagsservicegesellschaft mbH vertritt

Printed by Books on Demand GmbH, Norderstedt / Germany